Science and *Anti*-Science

Science and *Anti*-Science

by

Morris Goran, Chairman

Physical Science Department
Roosevelt University
Chicago, Illinois

ann arbor science PUBLISHERS INC.
POST OFFICE BOX 1425 • ANN ARBOR, MICHIGAN 48106

Copyright C 1974 by ANN ARBOR SCIENCE publishers, inc.
P.O. Box 1425, Ann Arbor, Michigan 48106

Library of Congress Catalog Card Number: 73-90416
ISBN: 0-250-40049-9

Manufactured in the United States of America
All Rights Reserved

for Cimarron and Toby

Other books by Morris Goran—

Introduction to the Physical Sciences, 1959
Experimental Chemistry for Boys, 1961
Experimental Biology for Boys, 1961
Experimental Astronautics, 1967
Experimental Earth Sciences, 1967
The Core of Physical Science, 1967
Experimental Chemistry, 1967
The Story of Fritz Haber, 1967
Biologia Experimental, 1967
The Future of Science, 1971

Preface

This is a book by a scientist about scientists and science. It will be read by scientists and nonscientists. For the latter it serves as a portrait painting by, say, Rembrandt; for the scientist it is a mirror. The overall conclusion one draws from this book is that scientists are, after all, human. They may average slightly higher in intelligence, curiosity, patience, creativity and objectivity but not much.

Most laymen form their opinions about science from a few impressions usually of a few scientists who find themselves newsworthy—Edison, Einstein, Salk, Pauling, Lysenko, for example. The result is not a very balanced impression. This book helps to even out the impressions by showing both good and bad, successes and failures, understanding and misunderstanding. As a result, the scientist emerges as a human being with all of the strengths and weaknesses of other sinners in this world.

Perhaps the worst sin of the scientist is nonobjectivity and Dr. Goran portrays many examples in many forms. A scientist can fall in love with his hypothesis and spend his life trying to prove its correctness rather than trying to objectively learn whether it is true or not. Prime examples are Lysenko and the inheritance of acquired characteristics, Shockley and the IQ of Blacks and many other causes célèbres—spontaneous generation of biological life, flat earth, perpetual motion, transmutation of base metals into gold, solacentric *vs* terracentric solar system, UFO's etc. Dr. Goran helps us to see how these situations arise because the scientist is human. He is just as emotional, defensive and greedy as school teachers, ministers or reporters.

This is a thoughtful and thought-provoking book written by a scientist who has taken time to think about the role of the scientist in yesterday's evolution into today's technocracy. His research into the ways of science has brought out hundreds of fascinating examples from the lives of hundreds of scientists from Democritus to Watson. He has contributed a book that will help the scientist understand himself and the rest of the world understand the scientist.

Walter C. McCrone

Contents

PART ONE   CONFLICTS

1. Conflicts with Governments 3
2. Conflicts with Religion....................................... 15
3. Conflicts with Intellectuals 23
4. Conflicts about the Nature and Conduct of Science 31
5. Conflicts about Subject Matter............................... 37
6. Personality Conflicts in Science............................. 45

PART TWO   MYTHS

7. Myths about Scientists....................................... 51
8. More Myths about Scientists................................. 59
9. Pure and Applied Science 67
10. Certainty in Science 77

PART THREE   SURPRISES

11. Surprises about Scientists 85
12. Surprises about Science 95

PART FOUR   MISLEADINGS

13. Misleadings by Organizations of Scientists 105
14. Misleading by Individual Scientists 111

    Index 117

Introduction

The nature of the scientific enterprise merits as much examination as does the nature of business, politics, education, and religion. Scientific and technological activity has, at least, an equal impact on individual lives. The influence has been noted by retired scientists as well as those at leisure, educators, philosophers, theologians, and men and women of affairs, but analyses have been minute. Until recently, the comments have been laudatory and the grumblings muted. Now the topic has become worthy of dispassionate scrutiny.

This book is concerned with the ways of science in four usually overlooked areas. First are its external and internal conflicts, that is, the disputes between science and other agencies and disciplines as well as the arguments within the family. Next are some of the myths about science and scientists. Two of these, the meaning of pure and applied science and the degree of certainty in the subject, merit special consideration. The third section, concerned with surprising material about science and its practitioners, is also related to the myths. Finally, misleading efforts by scientific or quasi-scientific organizations and by individual scientists are highlighted.

All of the topics are concerned with the truth about science but are also involved with anti-science. From the start, false ideas were circulated and adversaries to the scientific enterprise came from many quarters. This book deals with the nature of science and its opposition.

PART ONE

Conflicts

After being almost universally acclaimed at the beginning of the eighteenth century, Isaac Newton proclaimed that he stood on the shoulders of giants; he gave one of the tenets of cooperation between scientists: Prior art is freely available to all investigators. During the twentieth century, Ernest Rutherford expressed the concept another way: "It is not in the nature of things for any one man to make a sudden violent discovery; science goes step by step, and every man depends on the work of his predecessors . . . Scientists are not dependent on the idea of a single man, but on the combined wisdom of thousands of men." When Charles Darwin and Alfred Russel Wallace had a gracious relationship even though both were on the verge of announcing a theory of natural selection for organic evolution, they showed how men and women of science are able to cooperate for the advancement of knowledge. The international cooperation of scientists has also been cited as a prototype for other disciplines, notably politics.

Science has inevitably been presented as a united effort by the people of the laboratory. True in one sense, this portrait neglects another vital aspect of the activity — conflict, the other side of the coin of cooperation.

CHAPTER 1

Conflicts with Governments

Scientific research has been supported by industry, universities, associations and individuals, but by far the most extensive financing has been through governments. At the birth of modern science, Tycho Brahe had a lavish island observatory through the courtesy of the Danish king; when that munificent subsidy ceased, Brahe lived well at the court of Rudolf II of Bohemia. When the Académie des Sciences was founded in 1666, its members, including foreigners such as astronomers Cassini and Roemer, received salaries from the state. In the succeeding centuries, Western European governments, except for the British, gave money for scientific work; the amount was not large because scientific activity was not extensive. Money for research exploded in our time because governments began to view science as an ally in war and peace. In the United States, the federal government's share of total research and development expense rose from 25% in 1946-47 to 70% in 1959-60.

The twentieth-century reliance of governments upon science and technology has not prevented the occasional suppression of science by large powerful states. In some cases, the opposition saw science as a threat because of a particular result of objective investigation. The method, spirit, and personnel of the laboratory was the target at times. In several instances, unnecessary obstacles have been placed before scientists visiting foreign countries. In 1966, a French woman biochemist invited to participate in a New England meeting had to certify, in accordance with the U.S. McCarran-Walter Act, that she was neither a prostitute nor a Communist.

When scientists as citizens criticize a government, the government's response is invariably against the individual, not science. A case in point is that involving mathematician Stephen Smale. In 1965, he was chairman of the University of California's Vietnam Day Committee. In 1966, he received the prestigious Fields Medal at the International Congress of Mathematicians for his work in differential topology. At a press conference in Moscow, he said that Soviet intellectuals were unhappy about being unable to speak out as he was. Smale compared U.S. involvement in Vietnam to Soviet intervention in Hungary. He was summoned from the press conference by Soviet authorities but later returned to the mathematics meeting. The *New York Times* reported Smale saying, "It seemed to be a rather rude attempt to keep me from talking with Western correspondents." His federal government grant was later threatened, but a

3

concerted effort by the scientific community prevented any damage.

Another successful group action against an arbitrary governmental edict occurred during the early years of the Eisenhower administration. A company making an automobile battery additive, ADX2, wanted the National Bureau of Standards to give the product a seal of approval. The director of the bureau, Allen V. Astin, refused on the grounds that the Bureau of Standards had no such mission. Because the manufacturer had much political influence, Astin was forced to resign his post. The furor of objection in the scientific and technical community, abetted by responsible men of affairs, was so great that Astin was reinstated within a very short period.

The situation of J. Robert Oppenheimer, also during the Eisenhower administration, was not similarly handled. Oppenheimer was denied access to classified information despite his yeoman efforts in helping to build the atomic bomb during World War II. A three-man panel heard testimony, and by a two-to-one decision, they found him guilty of poor judgment with respect to Communist associations and referred to "fundamental defects in his character."

In several other twentieth-century cases, group action by the scientific community was to no avail. Anti-science movements by governments in Germany, South America, the U.S.S.R., and the United States proved to be too powerful.

The Nazis in Germany during the 1930s devastated the country's scientific potential because of their tirade against Jews. Expert non-Aryans emigrated, and every branch of science and technology was adversely affected. Feeble opposition by a few leading scientists and engineers was useless. Max Planck was silenced by Hitler's threats. A scientist of noble Aryan heritage, after consulting with his friends and family, proclaimed, "We cannot draw our swords for the Jews." Engineer Carl Bosch tried to launch a movement to protect Jewish scientists, but all his efforts resulted only in his sheltering of Hartmut Kallman in an industrial laboratory.

The contention that the Nazi onslaught was directed against individuals rather than science is difficult to support. Their racial doctrines had no objective standing, and the labeling of internationally accepted results as "Jewish" physics or "Jewish" biology was contrary to the spirit and letter of the ethics of the laboratory. Then, too, the dismantling of German science through the removal of Jewish personnel was condemned in other countries by disinterested observers.

Another argument is that German science could not have been seriously impaired because, after the country's collapse in World War II, the Allies diligently competed for its technical personnel. The United States persuaded Wernher von Braun and his team to emigrate, and ex-German scientists were cited by the envious when Russia had its rocketry successes beginning in the 1950s. Yet the Germans could not get their atomic bomb project going during World War II, and a host of refugees from the Nazis were involved in the Allied success with the weapon.

In the 1960s, Argentina also decimated its scientific potential. The Ongania dictatorship was apparently intent upon squelching all dissent, and the scientific community seemed to be a potent source. Stormtrooper tactics were used, and several laboratory workers were injured. About two hundred of the three hundred and fifty members of the University of Buenos Aires science faculty resigned. By 1967, the physics, inorganic chemistry, and applied mathematics departments were not operating.

In 1968, two prominent Mexican philosophers of science, Eli de Gortori and Nicolas Molina Flores, were arrested and held without trial for two years. Early in April 1970, the Brazilian government forced ten biologists at the Institute Oswaldo Cruz, well known for research in tropical medicine and hygiene, to resign and took away their "political rights."

In the U.S.S.R., the conflict between science and government was disastrous to genetics and no doubt affected the sciences peripheral to the subject. The government supported the claims of the mountebank Lysenko, who proclaimed that acquired characteristics are inherited. Those who openly disbelieved were often exiled to Siberia although the current Soviet tactic is to incarcerate the "offender" in an institution for the mentally unbalanced. Lysenko reigned over genetics throughout the 1950s and the early 1960s, a time long enough for the virtual destruction of the science in the Soviet Union.

In 1972, authorities of the U.S.S.R. made life miserable for their scientists of Jewish ancestry who wished to migrate to Israel. The most prominent to suffer the harassment, electrochemist Veniamin G. Levich wrote: "The international scientific community, I believe, should not consider such problems as a personal affair of each of us, but as a problem of the professional honour, dignity, and humanism of all the scientists all over the world."

Only three scientific organizations in the United States protested the Soviet position. At its meeting in late October 1972, the American Physical Society, with 28,000 members, approved a statement proposed by the Society's Forum on Physics and Society. The Federation of American Scientists adopted a resolution opposing the Soviet stand in general terms. At its annual meeting in October 1972, the 2300-member Society for Neuroscience told the Soviets about the "dismay at the reported imposition of special and exorbitant emigration taxes upon scholars and scientists. No amount of rationalization will persuade the international scientific community that this is other than a discriminatory and oppressive policy which abrogates the universality of knowledge as well as the fundamental rights of human beings."

On a more private scale, a letter of protest written by Jacob Bigeleisen, a chemist at Rochester University, was circulated to the 950 members of the National Academy of Sciences and signed by about half of them before being sent to Moscow. Another letter written by Oscar Zariski, a Harvard University mathematician, was signed by 42 mathematician members of the National Academy of Sciences before being sent to the president of the Academy of Sciences of the U.S.S.R., M.V. Keldysh, a mathematician.

In September 1973, the National Academy of Sciences joined in an international campaign of support for outspoken physicist Andrei Sakharov. Referring to official threats that Sakharov might be prosecuted for his criticism of Soviet government policies, the president of the National Academy of Sciences, Dr. Philip Handler, wrote to Dr. M.V. Keldysh: "Were Sakharov to be deprived of his opportunity to serve the Soviet people and humanity, it would be extremely difficult to imagine successful fulfillment of American pledges of binational scientific cooperation."

In hearings before the U.S. Senate Internal Security Subcommittee in 1972, Alexander Sergeyevich Yesenin-Volpin, a Soviet mathematician then at New York State University in Buffalo, told of having been confined in Soviet mental institutions five times since 1949. His last confinement came in 1968 because he had accepted an invitation to attend a conference in the United States. Only public pressure brought his release; 95 Soviet scientists signed and sent a letter to the Minister of Health in the U.S.S.R., calling the hospital detention of a talented scientist fully capable of work a flagrant violation of medical and legal norms. Yesenin-Volpin said, "If the civilized world protested in a louder and more courageous voice, we could bring about the liberation of far more victims of the Soviet political-psychiatric terror."

The Soviets used other weapons besides psychiatric incarceration for their dissident scientists. In 1972 the passport of physicist Valery Chalidze was seized while he was in New York. In 1973 the U.S.S.R. revoked the citizenship of gerontologist Zhores A. Medvedev while he was in London.

The conflict between science and government in the United States has taken a greater variety of forms and without the physical torture or death prevalent in other countries. During the French revolution, chemist Antoine Lavoisier was beheaded, and one exuberant arresting officer proclaimed that the state had no need for scientists. At about the same time, chemist Ignatius Martinovics was beheaded in Budapest for trying to spread the ideas of the French revolution.

Individual states in America were at first pockets of resistance to the idea of organic evolution. Several enacted legislation prohibiting instruction in the theory. At the 1925 Scopes trial in Dayton, Tennessee, one such law was effectively challenged through the courtroom expertise of attorney Clarence Darrow. Yet, as late as 1970, Mississippi had an anti-evolution law, voided in December 1970 by an appeal court. Later, the federal government, although an unrivaled patron of science, came into conflict with the discipline on at least two important issues. One was adequate funding and the second was reluctance of the establishment to accept scientific truth.

Science in the United States began to be well financed by the government at the start of World War II. The Soviet Sputniks brought an even greater U.S. expenditure of public money for science and engineering. Even before the first moon landing on July 20, 1969, the trend had already reversed. In February 1968, the American Institute of Physics claimed that 16% of academic physicists had lost all federal support. In face of costs of the Vietnam War and pressures for social welfare programs, science tumbled in priority; the space program was severely pruned, the Department of Defense virtually halted its underwriting of research vastly peripheral to security, and the National Science Foundation was not given money in the style to which it had been accustomed.

The attitude of Congressional leaders toward funding science became more skeptical and is indicated by the remarks of Rep. Craig Hosmer of California in a late 1971 hearing on the status and future of controlled thermonuclear fusion. At one point he told a testifying scientist: "We are not doing this thing for kicks, Doctor, PhD kicks or otherwise. This is something we hope to evaluate on a scale of necessity to the world and consider every single bit of money."

Scientists, spokesmen for the discipline, and assorted friends did speak for restoration of public money, but the timing of the appeals was unfortunate. A crescendo of criticism against pollution associated science and engineering with the malady, and the furor against bombs, napalm, and defoliation associated science with war. The public seemed to become apathetic to successive television spectaculars of moon landings in the midst of increasing crime, racial tensions, and political corruption. The call was for social and humane projects, not scientific research. Solutions to practical problems were called for, and lesser funds were provided for arriving at those solutions.

The scientific enterprise did have some friends in Congress. Referring to threats to the continuation of some National Institutes of Health programs, Rep. William R. Roy of Kansas, a member of the House health subcommittee, said, "any jackass can kick down a barn but it takes a carpenter to build one."

The fiscal conflict was visible and audible in contrast to another, more recent argument with governmental forces. In several instances, the government marshaled its own "experts" to oppose a scientific proposition. The opinions and conclusions of panels of scientists were ignored or belittled.

A group at Stanford University studied six questions related to the environment and public health and found, in 1970, that the conclusions of the scientific advisory groups were never followed. In the cases of supersonic transports, cyclamates, nuclear power plants, underground nuclear tests, pesticides, and herbicide used in Vietnam, the government virtually laid aside the recommendations. In spite of a strong disapproval of the SST, the administration appointing the advisory committee continued with enthusiasm to support the project for supersonic aircraft. The report by the Atomic Energy Commission's Advisory Committee on Reactor Safeguards did not guarantee the safety of a nuclear power plant at Lagoona Beach, Michigan; the document was suppressed and the installation did have a breakdown. A 1963 report urged that the use of DDT be drastically reduced; the reduction was finally accomplished by 1973.

Crusading scientists sometimes called public attention to the inertia of the establishment. In the battle over radiation standards, two University of California scientists, John W. Gofman and Arthur R. Tamplin, pitted themselves against such units as the Atomic Energy Commission and the Federal Radiation Council. Gofman and Tamplin charged harassment, while an AEC commissioner at a meeting sponsored by the Atomic Industrial Forum, in Los Angeles on February 9, 1970, said: "Gofman and Tamplin and their allies . . . are trying their case in the press and other public forums. We used to call such characters 'opera stars.'" Gofman and Tamplin urged a great reduction in allowable radiation exposure. Less strident and abrasive in their approach were scientist-critics of panels giving a virtual seal of approval to the continued use of the herbicide known as 2, 4, 5-T and of another panel claiming that lead in the air was not a serious hazard. In the first case, a strongly worded dissenting report was written by one member of the panel, and the Committee for Environmental Information as well as outstanding ecologists challenged the report; the final decision was to maintain the restrictions. The lead-in-air report was prepared by the National Research Council under its contract to the Environmental Protection Agency to review the literature on selected pollutants. Some environmentalists, including two who contributed to the document issued September 1971, questioned the inclusion of the lead industry representatives on the panel. Four of the eighteen authors were employed by either duPont or the Ethyl Corporation who, together, produce most of the lead additives burned in the United States.

Despite these challenges, a former Secretary of the Interior, Stewart

Udall, severely criticized scientists. He called them "Uncle Toms" who sold their souls to the Federal government.

At a Washington, D.C., conference in January 1972, arranged by the Center for Science in the Public Interest (a small nonprofit group), a molecular biologist from the Massachusetts Institute of Technology more or less summarized the conflict between science and government in the United States. David Baltimore said: "American science is rapidly becoming a state science." He warned that politicians are making the decisions on what scientists should investigate and who should get money. "Unless the scientific community reacts soon, it will be too late to salvage the freedom which has allowed scientists to make significant contributions to society."

The government has already trespassed on a freedom which the scientific community dearly cherishes: the right to acquire information. During World War II, U.S. scientists chafed under severe secrecy provisions, and soon after the end of the war, they sought a return to old freedoms. At the University of Chicago, one leader, physicist Samuel K. Allison, said in a public speech that the good scientists would be found working on the color of butterfly wings unless security measures were relaxed.

The Freedom of Information Act became law on July 4, 1967. President Lyndon Johnson, signing the statute, said: "I have always believed that freedom of information is so vital that only the national security, not the desire of public officials or private citizens, should determine when it must be restricted." Yet when a researcher at Stanford University Medical School asked the Food and Drug Administration for data on food additives and pesticide residues, the reply, about a year later, was to ask the inquirer to be more specific. When he narrowed his request to sodium nitrate, he was told "toxicological and other technical information is valuable commercial property that is regarded as confidential information." He was offered some data for a compilation and copying fee of $99.50. A Washington, D.C., housewife asked for the toxicological data relating to birth control pills; she was told that some of the information was protected under the Freedom of Information Act. The reply to another inquirer stated that the requested material was mixed in another file and it would cost $12,600 to separate out the data in addition to a copying charge of $0.25 per page and an advance fee of $5,000 to be subtracted from the total bill.

However, the Environmental Protection Agency spelled out its procedures under the Freedom of Information Act. William D. Ruckelshaus, the administrator, said: "It is the policy of EPA to make the fullest possible disclosure of information to any person who requests information without unjustifiable expense or delay." According to the regulations published in the Federal Register on December 3, 1971, the Agency has up to thirty days to search for requested documents. Bona fide trade secrets are protected, but the time spent by the Agency in ruling on claims of trade secrecy would be at no cost to the party requesting the information. When the Stanford University researcher convinced the Environmental Defense Fund to file suit on his behalf in February 1972, the general counsel of the FDA wrote about one month later promising to release the data dealing with sodium nitrate and asking that the lawsuit be dropped. He wrote: "We are in the midst of adopting a new policy for releasing information to the public. It is regrettable that the lawsuit was necessary, but we can certainly understand your reasons for it." Another case tried in court involved the attempt by the Office of Science and Technology to suppress an adverse report on the supersonic transport plane. The judge ruled that the Freedom of Information Act "requires that the disclosure requirement be construed broadly, the exemptions narrowly."

The Freedom of Information Act has several loopholes exploited by governmental agencies. During March 1973, only a small portion of the 35 advisory committee meetings of the National Institutes of Health were open to the public. Lawyers at the Department of Health, Education, and Welfare cited two loopholes: the protection of trade secrets and "personnel and medical files and similar files, the disclosure of which would constitute a clearly unwarranted invasion of personal privacy."

The argument with government is essentially about who controls and who interprets. This is the issue in every country regardless of whether the real dispute is financing or if scientists should be on tap or on top. Since governments are the source of the bulk of funds for science, the dilemma is vexing for those who do not want to bite the hand that feeds them. Shall scientists accommodate their interests to the needs of political administrations, attempt to acquire power through democratic or other means, or challenge through direct onslaught by publicizing the truth?

Another possible solution, influencing governments via the administrators who are scientists through training or recreation, seems remote. The U.S.S.R. has a percentage of government officials first educated in engineering or science; the United States has some who have shifted their employment from the laboratory to central offices. In both cases, however, the loyalties of the people probably belong to the state. Even those in earlier generations who were amateur scientists *and* politically important would probably not be science oriented. Emperor Hirohito of Japan has published several papers in zoology; King Carlos I of Portugal was a pioneer in marine biology early in the twentieth century; John Winthrop, Jr., governor of Connecticut, was also a scientist and a charter member of the Royal Society; the Holy Roman Emperor Frederick

II of Hohenstaufen contributed to ornithology with a treatise on falconry and dabbled in macabre physiological experiments.

The Nixon administration in the United States knew the importance of political loyalty for its science administrators. For example, in 1969 an eleventh-hour shift quashed the appointment of Franklin Long to the National Science Foundation directorship because he was politically suspect. Other cases included section heads for the Food and Drug Administration and the National Science Foundation.

The option of compromise is alien to laboratory workers. The operation of nature has never been probed by wheeling and dealing. The halo of truth-gatherer would be severely tarnished for the scientist who chooses to suppress, eliminate, or alter his results. Only some administrators in science, long removed from doing research and with short memories for laboratory mores, have so strong a loyalty to political administrations that they can get themselves to compromise the truth.

The alternative of driving for scientist power is just as unlikely. The aim of investigating nature would have to be set aside for another motivation — politics — and the record of scientists as public politicians is dismal. As a member of Parliament, Isaac Newton never entered into debate. Napoleon said the Marquis de La Place brought the spirit of infinitesimals into the government. In Germany, Rudolf Virchow, the physiologist, supported sewer building but voted to ban the teaching of organic evolution. The United States has had a couple of science-trained Congressmen with undistinguished records.

At a Northwestern University Colloquium on "The Control of Science for Civil Needs," April 1971, Anthony Wedgwood Benn, former Minister of Technology for the United Kingdom, dangled the appeal of political action. He said: "Now, all of a sudden, people have awakened to the fact that science and technology are just the latest expression of power and that those who control them have become the new bosses, exactly as the feudal landlords who owned the land, or the capitalist pioneers who owned the factories, became the bosses of earlier generations. Ordinary people will not now be satisfied until they have got their hands on this power and have turned it to meet their needs."

Scientists made a tiny stab at using their power in 1967. An international advanced study institute in molecular biology scheduled for the Greek island of Spetsai was postponed for a year. A United States member of the organizing committee said that the April 21 army coup in Greece represented "a considerable barrier to holding a meeting dealing with science in a tradition of free and open discussion."

Within the United States, scientists have formed small groups such as Society for Social Responsibility in Science, Federation of American

Scientists, and The American Physicists Association which may grow to real influence.

The beckoning siren of mass media has captured some scientists who chose another option fighting for the truth. Here the pitfall is the failure to recognize the huge amounts of space and time necessary to counteract the media readily available to the establishment. Being heard on a television program aimed at the literate makes a small dent for the scientist when it is compared to the impact of vast exposure via newspapers, magazines, radio, and television that the government employs.

At the start, attempts at political and economic power did not have too much support. During the days when popular fronts against fascism were in vogue, scientists had groups called "associations of scientific workers" that were often maligned as being communist-sponsored. When unemployment began to surge in the United States during the late 1960s and the early 1970s, the scientific societies began to employ professional lobbyists in Washington. In 1972, the American Society of Mechanical Engineers, with more than 50,000 members, hired a former Air Force lieutenant colonel as its representative. The Institute of Electrical and Electronic Engineers, with over 150,000 members, employed a man for the same purpose at almost the same time. The Political Action Committee for Engineers and Scientists started in late 1971 and began to lobby in Congress.

The effort in Western Europe is comparable. Some of the attitude was expressed by Hubert Bloch, vice-president of the Swiss National Science Foundation, at the Ciba Foundation Symposium, "Civilization and Science." He said: "Since science has invaded the field of government, scientists have no choice but to take a more active part in the shaping of policy; they cannot shun this responsibility. To limit their role to that of advisers without a share of the responsibilities no longer seems permissible."

There are some scientists who view any overt action as jeopardizing their apolitical position. They consider science to be a tribunal for truth rather than a political advocate. For them, the laboratory has an exalted position, supported by all political parties. Whatever the controversy with government, either misunderstanding or ignorance is responsible and the fault will be righted by men of good will without the raucous intervention of people who should be immersed in their slide rules and test tubes.

The conflict with national governments can be handled another way, in line with the tradition of science. Already some scientists are urging more international cooperation. In a letter to the *Times of London* during the summer of 1966, several who helped organize the International Geophysical Year and the International Years of the Quiet Sun proposed an extension of these successful ventures. Within the United States, scientists who have visited and worked in politically off-limits countries such as Cuba, mainland China, North Korea, North Vietnam and the U.S.S.R. have invariably indicated the warmth of reception and degree of cooperation received; only rarely do the letters to the editors of scientific journals complain about the treatment obtained in a communist country.

Many adherents of modern science have considered themselves citizens of the world. Astronomer Tycho Brahe, who was forced to give up his position as master of the Danish island of Hven and migrated to the court of Rudolf of Bohemia, wrote, in 1597: "And when statesmen or others worry him too much then he should leave with his possessions — with a firm and steadfast mind, one should hold under all conditions, that everywhere the earth is below and the sky above, and to the energetic man, every region is his fatherland." French philosopher and scientist René Descartes did not hesitate to be in the employ of a Netherland general or Queen Christina of Sweden. Dutch physicist Christian Huyghens lived and worked in Paris while his country and the French were at war. England and France were at war when English scientist Humphry Davy said, upon receiving a medal from France: "Some people say I ought not to accept this prize, and there have been foolish paragraphs in the papers to that effect. But if two countries are at war, the men of science are not. That would indeed be a civil war of the worst description. We should rather through the instrumentality of science soften the asperities of national hostility."

The first international scientific conference — on marine meteorology — was held at Brussels in 1853. In 1860, the first international congress of chemists was held at Karlsruhe, Germany, with one hundred and forty delegates. Between 1930 and 1942, some five hundred international scientific and technical conferences were held, and the pace increased after World War II. In August 1955, more than twelve hundred delegates from seventy-two nations came together for two weeks at the International Conference on the Peaceful Uses of Atomic Energy. Later, a resolution of the United Nations Conference on Food and Agriculture said, "The natural sciences are a particularly fruitful field for international cooperation because they are themselves international; basic physical and biological laws are universally accepted."

In an address to an international microbiology conference in Stockholm in 1963, Abba Eban, then Deputy Prime Minister of Israel, expressed the international cooperation achievement of science: "You have achieved what is remote from us, the basis of a universal community of thought and action, respecting natural differences, but also transcending them."

The success of international cooperation in science stimulates the idea of international funding of the discipline. When money for research flows from an international authority, conflict with government is minimized or abolished. Plans for the establishment of an international science foundation were discussed at a meeting in Stockholm, July 1970. The conference was sponsored by the Royal Swedish Academy of Sciences, the United Nations Educational Scientific and Cultural Organization, and the American Academy of Arts and Sciences. An international foundation (or fund) for scientific and technical development was recommended at the 19th Pugwash Conference on Science and World Affairs at Sochi, U.S.S.R., in October 1969, as it had been at two previous Pugwash Conferences also.

SELECTED READINGS

Doty, Paul. "The Community of Science and the Search for Peace," *Science,* **173** (September 10, 1971), 998-1002.

Goran, Morris. *The Future of Science* (New York: Spartan Books, 1971), Chapter 3.

———"Technologists as Politicians," *Chemical Technology,* **2** (March 1972), 191-192.

——— "A New Role for UNESCO," *International Science and Technology,* **2** (June 1962).

———"Politics for Scientists and Engineers," *Industrial Research,* **6** (March 1964), 25-27.

———"International Funding Sought for Basic Research," *Industrial Research,* **10** (September 1969), 21.

———"Money for Pure Science," *The Chemist,* 49 (March 1972), 101-102.

Haberer, Joseph. *Politics and the Community of Science* (New York: Van Nostrand Reinhold, 1969).

Von Hippel, Frank and Joel Primack. "Public Interest Science," *Science,* **177** (September 29, 1972), 1166-1171.

Ladd, Everett C., Jr. and Seymour M. Lipset. "Politics of Academic Natural Scientists and Engineers," *Science,* **176** (June 9, 1972), 1091-1100.

Medvedev, Zhores A. *The Medvedev Papers. Fruitful Meetings between Scientists of the World; Secrecy of Correspondence Is Guaranteed by Law* (New York: St. Martin's Press, 1971).

Mellanby, K., Terence Price, and J.R. Ravetz. "Conflicts of Loyalty in Science," *Nature,* **234** (November 5, 1971), 17-21.

Penick, James L., Jr., Carroll W. Pursell, Jr., Morgan B. Sherwood, and Donald C. Swain, (Eds.). *The Politics of American Science, 1939 to the Present* (2d ed., Cambridge: MIT Press, 1972).

Reagan, Michael D. *Science and the Federal Patron* (New York: Oxford University Press, 1969).

CHAPTER 2

Conflicts with Religion

The challenges of science to religion irritate the nonscientist much more than do the relations between science and governments. The emotional attachments of people to the concepts of theology, whatever the denomination, are strong enough to thwart any questioning. Men and women seem to show greater loyalty to their religious faith than to their country.

The conflict between science and religion has been much more publicized than that between science and government. The first is a matter of history; the second, a current event, is central to the continued health and growth of science. The fights between science and religion have been well documented; those with governments have been largely neglected except for analyses by a small coterie of political scientists. Religion has a sordid record as a supporter of research, and no agency — industry, private patron, or university — has financed science as splendidly as have governments. Monarchy or democracy, despotic or liberal, the state has a use for science; religion does not gain sustenance from the subject. Compared to those with governments, the conflicts with religion have hardly affected the overall progress of science.

Early Christian commentaries appear to be opposed to the spirit of inquiry. During the fourth century, Saint Basil, in his *Hexaemeron,* wrote: "At all events let us prefer the simplicity of faith to the demonstrations of reason." Saint Ambrose is supposed to have said: "To discuss the nature and position of the Earth does not help us in our hope of the life to come." Saint Augustine is credited with: "The good Christian should beware of mathematicians and all those who make empty prophecies. The danger already exists that the mathematicians have made a covenant with the devil to darken the spirit and to confine man in the bonds of Hell."

Those reactions, however, are not unique to the early Christians. The outstanding Greek philosopher, Plato, has been accused of being antiscientific; in some circles, Aristotle is placed in the same camp. Moreover, the first Christians lived in the Rome-dominated society that did not nurture the Hellenic scientific and philosophic achievements. The decline of science in the western world started in the Roman empire.

Saint Augustine saw science and theology as two separate realms with a gulf between them; conflict was not in the system. He wanted his fellow Christians to know science so that the fundamental ideas of the religion would not be jeopardized by absurd statements. He wrote: "If we come to

15

read anything in the Holy Scripture that is in keeping with the faith in which we are steeped, capable of several meanings, we must not, by obstinately rushing in, so commit ourselves to any one of them that, when perhaps the truth is more thoroughly investigated, it rightly falls to the ground and we with it."

The miniscule amount of scientific activity in the first thousand years after Christ was not altered by Christianity. Indeed, the early leaders of Christianity could be cited as having opinions close to the modern scientific one on such a fundamentalist-rankling topic as the origin of life. Saint Augustine wrote: "God as a rule creates wine from water and earth through the mediation of grapes and their juice; however, sometimes, as in Cana of Galilee, he can create it directly from water. Thus also, in respect of living things, he may cause them to be born from seeds or to emerge from inanimate matter where invisible spiritual seeds [*occulta semina*] repose." A similar opinion can be found in the writings of Saint Basil, authority for the Eastern branch of the Church as Saint Augustine was for the Western.

Another way to look at the lack of conflict between early religion and the embryo of science is to cite the protection of learning sponsored by church groups. Monasteries cherished manuscripts and harbored scholars in theology. Even more significant is the advance of science occurring with the rise of Islam. For several centuries after Mohammed's death in the seventh century, the Moslem world had great activity in many branches of science. For example, Alhazen (965-1038) did pioneering work in optics, and Avicenna (980-1037) was a supreme authority in medicine for six hundred years.

Clerics fostered the germ of science prior to the Renaissance. In the twelfth century, the English Adelard of Bath investigated phenomena in an objective manner. He said: "I do not detract from God. Everything that is, is from him and because of him. But [nature] is not confused and without system and so far as human knowledge has progressed, it should be given a hearing. Only when it fails utterly should there be recourse to God." In the thirteenth century, the English Franciscan Roger Bacon claimed that God showed his mind in two ways, in the Scriptures and in nature; both should be studied, he said, but the latter had been neglected.

Albertus Magnus, who tried to reconcile Christian theology and Greek philosophy, and Moses Maimonides, who tried to do the same for Jewish thought, can be considered early scientists. The latter was the author of ten medical treatises; he strongly condemned astrology in a letter to the rabbis of Marseille in 1194. Robert Grosseteste, the Bishop Lincoln, was probably the first western writer after Aristotle to explore the meaning of scientific method. Peter of Spain, who became Pope John XXI, was also a physician who wrote about the value of experiment. Nicolas Oresme, the

Bishop of Lisieux, was a mathematician who translated Aristotle into French. And, of course, Nicolas Copernicus, the developer of the heliocentric system, was a Polish bishop.

At the beginning of the great squabble between Galileo and the Church, the latter had the services of the Jesuit astronomer Christopher Clavius, who soon died. Perhaps the outcome of the tiff with Galileo would have ended differently had Clavius been living. He was sensitive, enlightened, and receptive to the introduction of the telescope. The argument could have run another course, too, had the Church not been beleagured to maintain its authority in many countries and many realms. Corruptions, dissidents, and defaults weakened its structure, and its response was reprisal. As a loyal Catholic, Galileo was subject to the penalties for disbelief. His punishment was severe in modern intellectual terms but mild for a practicing member of the early Church.

Galileo had to give up publicly an idea he had conceived from observation and logical thought. Such a procedure is alien to the spirit of modern science. Evidence, not authority, dictates conclusions. Galileo was forced to agree to the following paragraphs:

> I wrote and printed a book which I discuss this new doctrine [that the earth is not the center of the world and moves] already condemned and adduce arguments of great cogency in its favor without presenting any solution of these. I have been pronounced by the Holy Office to be vehemently suspected of heresy, that is to say, of having held and believed that the Sun is the center of the world and immovable and that the Earth is not the center and moves.

> Therefore, desiring to remove from the minds of your Eminences, and of all faithful Christians, this vehement suspicion justly conceived against me, with sincere heart and unfeigned faith I abjure, curse, and detest the afore-said errors and heresies and generally every other error, heresy, and sect whatsoever contrary to the Holy Church, and I swear that in the future I will never again say or assent, verbally or in writing, anything that might furnish occasion for a similar suspicion regarding me.

About 100 years later in France, scientist-writer Georges-Louis Leclerc, Comte de Buffon, much better known in his lifetime than now, had to ascribe to a similar confession. He wrote: "I abandon whatever in my book concerns the formation of the earth, and in general all that might be contrary to the narration of Moses, having presented my hypothesis on the formation of the planets only as a pure philosophical speculation."

During Galileo's lifetime, extensive evidence in the modern sense was not available for the support of the Copernican doctrine. Those who took to the heliocentric view were propelled by criteria of simplicity and elegance, not observational data. Galileo, then, cannot be viewed as an injured party in the sense that he had support for the truth. He was hurt only in being made to go against his intellectual belief.

The Catholic heirarchy was not the only religious opponent of the sun-centered idea. The Lutherans did not accept the concept. Martin Luther said, in 1539: "People gave ear to an upstart astrologer who strove to show that the earth revolves, not the heavens or the firmament, the sun and the moon . . . this fool wishes to reverse the entire science of astronomy; but the sacred scripture tells us [Joshua 10:13] that Joshua commanded the sun to stand still, and not earth."

The opposition of organized religion did not appear to hinder the progress of science because the discipline grew immediately after the Galileo conflict. Likewise, after the second important fight between religion and science — about organic evolution — the laboratory had a spurt of activity.

The arguments did not prevent scientists from being religious or theologians from practicing science, true even to this day. A generation after Galileo, Isaac Newton, a genius in physics, considered his religious studies of prime importance. Blaise Pascal, in France, wavered between science and religion, finally being won by the latter. Robert Boyle was so religious that he would lower his eyes at every mention of God. In Holland, physician and teacher Herman Boerhaave refused to see Voltaire, who came especially to see him, because Boerhaave was not prepared to rise for someone who did not rise for God. In the United States, Cotton Mather was renowned as a cleric, and he was also a pioneer in plant hybridization and psychosomatic medicine. At the time of the American revolution, Joseph Priestley, a dissenting minister and well-known chemist, worried about other scientists losing their faith in an age of growing agnosticism; he wrote articles for "philosophical unbelievers," pointing out how un-philosophic it was to reject all the evidences for Christianity without proper examination. Living at about the same time, the founder of comparative anatomy, Georges Cuvier, was an ardent supporter of every word of the Book of Genesis in the Old Testament. During the nineteenth century, English physicist James Joule wrote that natural philosophy is second only to religion, and German physiologist Rudolf Virchow claimed the object of science is not to destroy faith. In the current century, English physicist J. J. Thomson said that every advance in science emphasizes that "great are the works of the Lord." Science administrator and mathematical physicist Warren Weaver said: "I am myself so old fashioned as to be deeply committed to organized religion." An excellent American nuclear physicist, Dr. E. Pollard is also a minister, while a group of Orthodox Jewish scientists have formed an association.

Today, the Church and science still occasionally have a dispute. At the 135th meeting of the American Association for the Advancement of Science (Dallas, Texas, December 1968), about two thousand scientists signed a document calling Pope Paul VI unenlightened. They termed his birth control encyclical "repugnant to mankind," charging him with promoting war and poverty and sanctioning "the deaths of countless numbers of human beings with his misguided and immoral encyclical." The Catholic Church has always been against birth control by any method other than so-called natural rhythm.

In 1970, an association of religiously fundamentalist scientists known as the Creation Research Society pressured the California legislature to consider a law whereby biology text books would present the development of life according to Genesis as well as the development described by modern science. The charter of the Society commits members to "full belief in the Biblical record of creation and early history, and thus to a concept of dynamic special creation (as opposed to evolution), both of the universe and the Earth with its complexity of living forms."

At their October 1972 meeting, the National Academy of Sciences adopted a resolution urging the California State Board of Education not to adopt a requirement that creationism be given equal prominence alongside the theory of evolution in school textbooks. The NAS claimed "The essential procedural foundations of science exclude appeal to supernatural causes as a concept not susceptible to validation by objective criteria; and religion and science are, therefore, separate and mutually exclusive realms of human thought whose presentation in the same context leads to misunderstanding of both scientific theory and religious belief."

NAS and other scientific bodies geared to oppose the California fundamentalists must also meet the attack in other parts of the United States. Members of the California Creation Research Society have been active in half a dozen other states. A poll taken September 1972, in the Dayton, Tennessee, high school showed 75% of the students believed the account in Genesis and not the concept of organic evolution. (It was in Dayton, Tennessee, in 1925 that attorney Clarence Darrow effectively ridiculed William Jennings Bryan and the fundamentalist idea in the trial of a teacher, Scopes, who had violated the state law against the teaching of evolution.) The Texas Board of Education requires all textbooks mentioning evolution to include in their preface a statement warning that evolution is a theory, not a fact. The religion editor of the *Washington Star-News* filed suit in Washington, D.C., to require biology textbooks to include the Biblical creation story.

The British counterpart of the Creation Research Society is the Evolution Protest Movement. Founded in 1932 by a submarine commander, its past presidents include Sir John Ambrose Fleming, the physicist who designed the first vacuum tube, and Sir Cecil Wakeley, who is also a past president of the Royal College of Surgeons.

The United States, too, has scientists, usually not biologists, who not only doubt the theory of organic evolution but also support the account in Genesis. When the weekly organ of the American Chemical Society, *Chemical and Engineering News*, reported the pressures on the California Board of Education concerning the Biblical story of creation, many readers bombarded the editors with support for the special creation concept. Most of the letters printed in the February 12, 1973, issue were so oriented. On the other hand, the seven letters about the topic published in the March 9, 1973, issue of *Science* were almost equally balanced; one writer wanted every Bible used in schools to have an explanation of Darwinism alongside the material in Genesis.

The idea of evolution was involved in an earlier major tangle between science and religion. In the last century, after Charles Darwin published *The Origin of Species*, the Church of England's Samuel Wilberforce made nasty remarks about the concept to educator and scientist Thomas Henry Huxley. In the eyes of most objective analysts, Huxley won the dispute. The level of the argument was individual because the Church of England made no official pronouncements against organic evolution. Only fundamentalist groups such as the Dutch Reformed Church in South Africa banned the teaching of evolution.

Save for fundamentalist sects, religion has more or less accepted science's descriptions of the place of the earth in the solar system and the development of life forms on earth. The past tensions have been reconciled. However, since new arguments are destined to arise, the lessons of past arguments should be learned by both parties.

Each, science and religion, has its own basic assumptions and procedures. The laboratory-oriented believe in the capabilities of man and have a fundamental faith in the rationality of the universe; those imbued with religion look to a higher authority than man for guidance and answers. Scientists at work do not ordinarily invoke the supernatural, mystical, or irrational; theologians have not much use for experiment and the quantitative approach.

Spokesmen for science as well as spokesmen for religion have commented — and often debated — about genetic engineering, first cause, life on other worlds, and a host of other topics, any one of which could explode into a confrontation such as engulfed the heliocentric conception or the theory of organic evolution. Ugly arguments could be avoided if both disciplines could offer their wares to the public in a free marketplace, minus pressures from individuals or groups. Since this ideal is transgressed by the fact of family, peer, school, neighborhood, and professional loyalties, educational institutions could try to overcome the prejudices one way or the other by presenting both subjects fairly and objectively. Through such a tactic, religion would gain school entrance in many countries. Since public examination of the issues is a tenet of science, it would gain expertise, now wanting, in presenting its case to the public.

SELECTED READINGS

Ellegord, Alvar. "Darwinism and Religion," in Richard Olson (Ed.), *Science as Metaphor* (Belmont, Cal.: Wadsworth Publishing, Co., 1971), 120-133.

Hiebert, Erwin.   "Thermodynamics and Religion," in Richard Olson (Ed.), *Science as Metaphor* (Belmont, Cal.: Wadsworth Publishing Co., 1971), 161-175.

Irvine, William.   *Apes, Angels, and Victorians: The Story of Darwin, Huxley, and Evolution* (New York: McGraw-Hill, 1955).

De Santillana, George.   *The Crime of Galileo* (Chicago: University of Chicago Press, 1955).

White, Andrew D.   *A History of the Warfare of Science with Theology in Christendom,* 2 vols. (New York: D. Appleton and Co., 1897).

Wood, John K. *The Nature of Conflicts between Science and Religion* (Logan: Utah State University, 1962).

Trinklein, Frederick E. (Ed.) *The God of Science* (Grand Rapids, Mich: William B. Eerdmans Pub. Co., 1972).

CHAPTER 3

Conflicts with Intellectuals

In 1966, the residents of the affluent Chicago suburb of South Barrington were opposed to the building of a large accelerator in their town, finally sited at Batavia, Illinois, because the influx of scientists would "disturb the moral fiber of the community." The residents of South Barrington were for the most part educated, yet viewed the scientific enterprise as alien. During November, 1968, the U.S. oceanographic research vessel *Silas Bent* was invited to Maizuru, Japan, by the Oceanographical Society of Japan. A group of about fifty students at Kyoto University's faculty of fisheries, also at Maizuru, protested; the Society canceled its invitation. The objecting students were not those in the humanities and social studies; they were majoring in an aspect of science.

One of the earliest recorded criticisms of the ways of science was of Thales of Miletus, known later as one of the seven sages of Greece. Thales predicted the solar eclipse of 585 BC, used the concept of conditioned reflex, and expressed the idea of parsimony. Observing the stars, he fell into a ditch, and an old woman attending him said: "How can you know what is doing in the heavens when you don't know what is at your feet?" Such belittling, in one form or another, has continued to the present day.

The literati usually were snipers, aiming at one feature or another of the scientific enterprise. A popular author in the France of his day, Bernarden de Saint Pierre rejected the methods of scientists as too analytical. Their procedures, he claimed in a book published in 1784, blinded them to the real beauty and harmony in nature. Author Samuel Butler accused the fledgling experimenters in the young Royal Society of being interested in "the braying of an ass"; poet Walt Whitman took offense at the astronomer's objective study of stars; and Victorian novelist George Gissing wrote:

> I hate and fear science because of my conviction that for a long time to come if not forever, it will be the remorseless enemy of mankind. I see it destroying all simplicity and gentleness of life; all beauty of the world; I see it restoring barbarism under the mask of civilization; I see it darkening men's minds and hardening their hearts; I see it bringing a time of vast conflicts which will pale into insignificance 'the thousand wars of old' and, as likely as not, will whelm all the laborious advances of mankind in blood-drenched chaos.

Current literati may be equally damning. According to Catholic philosopher Jacques Maritain, science set off the "deadly disease" of "the denial of eternal truth and absolute values." Swiss playwright Friedrich Duerrenmatt in his *The Physicists* portrays scientists as a callous group in a race for the power to destroy the earth. Philosopher Herbert Marcuse believes that the dehumanizing aspects of modern society are "inherent in pure science." Better known writers are more reserved in their criticism. Joseph Wood Krutch claimed: "We are disillusioned with the laboratory not because we have lost faith in the truth of its findings but because we have lost faith in the power of those findings to help us generally as we had once hoped they might help."

Immoral Science was portrayed by Charles Dickens when he wrote about a geologist who knocked samples off buildings; arrested for defacing property, the geologist said that he knew no building but the temple of science. Later, Hitler's henchman, Albert Speer, wrote in his memoirs: "Because of what seems to be the moral neutrality of technology, these people were without any scruples about their activities. The more technical the world imposed on us by the war, the more dangerous was this indifference of the technician to the direct consequences of his anonymous activities." Poet and novelist Robert Graves, writing in the British journal, *New Scientist* (December 12, 1971) said: "The worst that one can say about modern science is that it lacks a unified conscience, or at least that it has been forced to accept the power of Mammon. Mammon—or at least the Talmudic Mammon of Unrighteousness exploits the discoveries of science for the benefit of international financiers, enabling them to amass more and more money and, it is hoped, to control all markets and governments everywhere." And still another view of immoral science is that of Alexander Solzhenitsyn, who, had he been allowed to receive the 1970 Nobel Prize for literature, would have said:

> It would seem that the appearance of the contemporary world rests solely in the hands of scientists: All mankind's technical steps are determined by them. It would seem that it is precisely on the international good will of scientists, and not of politicians, that the direction of the world would depend. All the more so since the example of the few shows how much could be achieved were they all to pull together. But no: Scientists have not manifested any clear attempt to become an important, independently active force of mankind. They spend entire congresses in renouncing the suffering of others: Better to stay within the precincts of science. That same spirit of Munich has spread above them its enfeebling wings.

Scientists have a long history of working for an employer without raising the question of morality. The greatest scientist of antiquity, Archimedes, followed his king and colony, Syracuse, into the camp of Carthage, perhaps because Hannibal was defeating the Romans at Cannae, or maybe the new Roman king of Syracuse was bribed into joining the enemy. For three years, the Roman fleet under Marcellus was held at bay because of the ingenious devices invented by Archimedes. And before Galileo achieved renown, he offered to sell his telescope as a military weapon to at least two groups.

Revolutionary science was the charge at McGill University at the beginning of the century. Ernest Rutherford was uncovering the nature of the atomic nucleus. Several times, his colleagues in other departments of the university told him that the radical ideas he was proposing about matter might bring discredit to the university. He was asked to delay publication and proceed more cautiously.

Later, when the economy began to creak and a depression occurred, science and technology became easy scapegoats. There was a widespread suggestion that automation had brought on unemployment, and a moratorium for science and invention was proposed.

Conceited science is in a wide spectrum of places. A member of the American Medical Association Research Institute told a reporter in 1967: "As a group we never really fit in. The scientist in a lab coat and sandals and turtleneck sweater was a source of puzzlement to the AMA rank and file." Another, a Nobel prize winner, said: "They don't understand what we prima donnas in science are like." More often, conceited science is precipitated when chauvinistic scientists mouth their belief that scientific activity is the harbinger of good things for the good life and be-all and end-all of existence. A Russian biochemist, Vladimir Engelhardt, en- visioned that by the year 2000 mankind will have pep pills with no side and after effects, cancer will be conquered, and human organs will be transplanted with ease. Robert Sinsheiner, professor of biophysics at the California Institute of Technology, said, at the Conference celebrating the 75th anniversary of his institution, that human intelligence would soon be directing human evolution. Literati such as Joseph Wood Krutch in the January 20, 1968, issue of *Saturday Review* scoff at such arrogance. More recently, scientists have been proclaiming their methodology will help solve social problems; some nonscientists are aghast at this concept.

Objective science has been scored by Theodore Roszak in his *The Making of a Counter-Culture*. He charges: "While the arts and literature of our time tell us with ever more desperation that the disease from which our age is dying is that of alienation, the sciences, in their relentless pursuit of objectivity, raise alienation to its apotheosis as our only means of achieving a valid relationship to reality."

A less sophisticated view of the pitfalls of objectivity was described by columnist William F. Buckley in his article entitled "Scientific Objectivity Yields to Twaddle," November 23, 1971. Buckley complained that scientists "zoom about like Hell's Angels in the world of controversy" and are partisan rather than objective.

Impractical science is the target of men and women not usually classified as intellectual. The Russian novelist and philosopher, Leo Tolstoy (1828-1910), could be the spokesman for this group with his: "He expects that science will teach him how to live, how to act towards members of his family, towards his neighbors, towards foreigners, how to battle with his passions, in what he should or should not believe, and much more. And what does our science tell him concerning all these questions."

Scientists can be found in this camp. A chemist in the Department of Soils, Water, and Engineering, University of Arizona, showed the position in his letter to *Chemical and Engineering News* (May 7, 1973), writing: "What we need is to turn our attention as individuals from using science for increasing the quantity of our possessions; to redirect it toward improving our emotional environment so that we can survive as men (not machines) in an irrational world. What we need is not more light but more warmth."

Destructive science has long been a theme of those who cite the horrors of war, and lately protectors of the environment have joined the attack. Typical of the condemnation is that of Dr. Jacques Piccard, a scientist: "This technology we 'enjoy' today is little else but a widespread suicidal pollution. It is a blight affecting not only the air we breathe and the water we drink, but also the land we till and the outer space we hardly know. But most tragic of all, we now have the pollution of man in his body by insidious chemical products. Technology is working against Man, Man is working Nature, and instead of natural selection, only technology remains."

The view of the universe depicted by science is also not to the liking of the intellectuals. They are disillusioned and upset by the cold and impersonal world described by Bertrand Russell:

> Even more purposeless, more void of meaning, is the world which Science presents for our belief. Amid such a world, if anywhere, our ideals must henceforward find a home. That man is the product of causes which had no prevision of the end they were achieving; that his origin, his hopes and fears, his loves and his beliefs, are but the outcome of accidental collocations of atoms; that no fire, no heroism, no intensity of thought and feeling, can preserve an individual life beyond the grave; that all the labors of the ages, all the devotion, all the inspiration, all the noonday brightness of human genius are destined to extinction

in the vast death of the solar system and that the whole temple of man's achievement must inevitably be buried beneath the debris of a universe in ruins.

A similar view has been stated by Anglo-American philosopher W. T. Stace: "If the scheme of things is purposeless and meaningless, then the life of man is purposeless and meaningless too. Everything is futile, all effort is in the end worthless. A man may, of course, still pursue disconnected ends, money, fame, art, science, and may gain pleasure from them. But his life is hollow at the center. Hence the dissatisfied, disillusioned, restless, spirit of modern man."

Scientists are more perturbed now about the charges against their subject than they ever have been. At the modern birth of the discipline, intellectuals were allies of scientists, and agriculture, industries, and governments found applications of science to be worthwhile. One of science's earliest advocates, Francis Bacon, was an attorney and politican. The amateurs who formed and promoted the Royal Society were physicians and gentlemen of leisure. Intellectuals were the bulwark of the members in the scientific societies of France, Germany, Holland, and Italy. Poets celebrated the genius of Isaac Newton, and writers during the Age of Reason paid homage to the mind of man and its accomplishments as seen in Science.

Scientists are more concerned now because they realize that winning friends and influencing people is a necessity for continual funding. Patrons and industries cannot support research in a modern style; only public tax money can do the job. Government money requires public confidence for the fountain of wealth to be in steady operation. Public financing inevitably responds to the whims, fancies, and pressures of the electorate, and the scientific community has come to realize the situation.

In his Sigma Xi lecture at the University of Houston, October 21, 1970, Philip Handler, president of the National Academy of Sciences, more or less recognized the importance of public opinion. He pleaded eloquently: "If you believe—as I do—that science remains the most powerful tool the mind of man has yet conceived to alleviate the condition of his fellows— please say so. If you believe that the pursuit of science is not merely the expensive hobby of scientists but both the leading edge of our culture and the only rational basis for a better way of life tomorrow—please say so."

Some scientists and their organizations have made a frontal attack against the indictments of science. They show instances of scientific morality, tradition and conservatism, humility, subjectivity, and peaceful nature. This kind of breast-beating has made few, if any, converts among the contrary-minded intellectuals. For one, the scientist's journals, where the defense generally appears, are not read by the literati. Second and more important, the opposition to science is not swayed by reason.

A few decades ago, Edwin Grant Conklin in an address as retiring president of the American Association for the Advancement of Science said: "The fact is well attested that science has given us grander and more inspiring concepts of the universe, of the order of nature, of the wonderful progress than were ever dreamed of in prescientific times. And as an educational discipline there are no other studies that distinguish so sharply truth from error, evidence from opinion, reason from emotion; none that teach a greater reverence for truth nor inspire more laborious and persistent search for it." The number of nonscientists who listened and reflected was infinitesimal. Very few intellectuals are aware that the early twentieth-century mathematical physicist Henri Poincare said: "The scientist does not study nature because it is useful; he studies it because he delights in it, and he delights in it because it is beautiful. If nature were not beautiful, it would not be worth knowing, and if nature were not worth knowing, life would not be worth living."

Another approach is more philosophical and views the current opposition as a temporary phase of disenchantment. The cycle of reason and romance, appealing to the arm-chair strategist, demands no action. Those persuaded by this interpretation point to the infatuation with mysticism, the supernatural, and the pleasures of the body as temporary sidepaths until mankind again takes up the road of reason through science. A related view notes that every society has had its scapegoats, from evil spirits in primitive groups to outmoded institutions during the eighteenth-century Enlightenment. Now, goes the argument, science and technology are blamed.

From the start, reason and experiment have had a difficult time. The perspective of organic evolution indicates that a nervous system and its appendages, necessary for thought, came late in the development of life. Once established in more specialized animals, the mechanism was in continuous struggle with other, less flexible means to cope with the environment. In man, too, thinking was difficult, often painful, and sometimes at odds with other approaches. Nonetheless, it triumphed in enough situations to bring man to his present state.

Scientists are also in the camp of those who wish to bring the rule of reason, and presumably science, through education; yet, organized higher education is as beleagured as the laboratory. The indictment of colleges and universities is even more drastic. The Newman Report, issued by a task force called together by the Department of Health, Education, and Welfare, charges:

> All too infrequently is an undergraduate course organized or taught on the assumption that students might learn best through subjective or practical experience. . . . Rarely are there politicians or lawyers in political science departments, novelists,

clergymen, or practicing psychiatrists in psychology departments, or engineers asked to help teach courses in the department of physics. . . . Moreover, seldom do the majority of faculty
members spend any time in jobs outside the university.

Technology, too, is pointed out as a solution to the crisis. Its products,
like education's, are not always exemplary, but both processes have extensively modified the environment. Comparable to the critics who say
society would have been better off without so many educated are those who
look with nostalgia to a pre-industrial time. The technological remedy for
the unfriendliness of intellectuals, however, is in the realm of biological
engineering. Advances in molecular genetics would give the ability to mold
human beings, and progress in chemotherapy would bring almost total
control of mood, temperament, and even belief.

The vision of biological tinkering infuriates the intellectual. The human
body, especially the psyche, is inviolate to them. One certain way to make
them less friendly to science is to produce the vision of mortal man
tampering with our biological inheritance. What needs to be done is to
meet the intelligent opponent of science on his own terms and in his own
ground. No better delegate to the supposedly enemy camp exists than the
many scientists who are artists in their spare time (see Chapter 7). At the
same time, the scientific community must be more knowledgeable of the
entire spectrum of science rather than equate one segment of activity with
the subject. If some opponents would rather have their emotions stirred
than their brains ransacked, then the many places in science where strong
feelings arise, such as creation and conflict of ideas, should be cited. If
consumer conveniences are distasteful, then the beauty and grandeur of
scientific ideas should be stressed. If numbers and mathematics are a
strange language, then words and pictures can be employed.
Simultaneously, scientists may seize the opportunity to listen and learn.

SELECTED READINGS

Barzun, Jacques. *Science: The Glorious Entertainment* (New York: Harper and Row, 1964).

Bronowski, J. "Technology and Culture in Evolution," *The American Scholar,* 41 (Spring 1972), 197-211.

Brooks, Harvey. "Can Science Survive in the Modern Age?" *Science.* **174** (October 1, 1971), 21-30.

Cotgrove, Stephen. "Anti-Science," *New Scientist,* **59** (July 12, 1973), 82-84.

Goran, M. "The Literati Revolt against Science," *Philosophy of Science,* **7** (July 1940), 379-384.

McDermott, John. "Technology: The Opiate of the Intellectuals," *New York Review of Books* (July 31, 1969), 25-35.

Olson, Richard (Ed.). *Science as Metaphor, The Historical Role of Scientific Theories in Forming Western Culture* (Belmont, Cal.: Wadsworth Publishing Co., 1971).

Roszak, Theodore. *The Making of a Counter-Culture* (New York: Doubleday, 1968).

Snow, Sir Charles Percy. *The Two Cultures and A Second Look* (New York: New American Library, 1964).

Teich, Albert H. (Ed.). *Technology and Man's Future* (New York: St. Martin's Press, 1972).

CHAPTER 4

Conflicts about the Nature and Conduct of Science

The nineteenth-century English writer, Matthew Arnold, gave the early Greeks credit for discovering science, but almost every renowned thinker of that age has a varying perspective of the subject. For examples, Heraclitus, who lived about 500 BC, thought reason was inherent in nature; Anaxagoras, about a century later, took reason as one of the infinity of elements which go to make up the world. The scientific pioneering of Pythagoras is mixed with numerology, mysticism, and religion. Plato, the first philosopher whose opinions are more clearly known, gives indications of being hostile to astronomical observation and adopting an organismic approach; nonetheless, he is a man of science. Aristotle, an expert biologist, formalized logic and outlined a method for acquiring reliable knowledge; yet his ideas have been cited as a barrier to the progress of science.

At the birth of modern science, two leaders of the subject, a Frenchman and an Englishman, René Descartes and Francis Bacon, respectively, published different views about the nature of scientific procedures. Bacon more or less described a variety of induction, establishing generalizations from particulars; Descartes was an advocate of deduction, the application of general statements. Bacon was not a practicing scientist, yet he won accolades from later generations, mostly British, for showing the proper road. Descartes had scientific achievements but not nearly so many testimonials from thinkers after his time. Both men failed to see the entire picture; both failed to emphasize or overestimate the role of hypotheses, experiments and mathematics. Concomitant with the economic and political rivalry between France and England, each man had his devotees, interpreters, and misinterpreters. The ways of science came to be described in Bacon's or Descartes's manner until Isaac Newton dominated the scene.

The method of science according to Newton was challenged in his time. The great Englishman insisted *"Hypotheses non fingo"*—"I frame no hypotheses"—despite the fact that he did, and critics were confounded. Newton and his contemporaries as well as those before them agreed, however, that the value of science was in its support of theology. He considered his scientific work a minor part of his life compared to his dabbling in Biblical chronology.

The scientists of the Age of Reason took a different stance with respect to being subservient to religion. Typical was the remark of Marquis de La Place when asked why he did not mention God in all his works. "I have no need of that hypothesis," was his curt reply.

The Age of Reason saw the birth of the controversy traceable to Plato and Aristotle and accentuated by Bacon and Descartes. According to philsopher Immanuel Kant, the mind of man has pre-established concepts with which to survey nature. In contrast, Newton's contemporary, John Locke, claimed our minds were like a blank slate on which our experiences with nature were written. In England in the nineteenth century, the debates between John Stuart Mill, representing the empiricist school, and William Whewell, for the other side, seemed to change nobody's position about the ways of science.

Today's arguments concerning the philosophy of science are often fierce and unrelenting. On occasion, discussions of positivism, the school with disdain for the metaphysical and the immeasureable, are replete with diatribe and contumely and are, for the distant observer, not far from fisticuffs.

Scientists not tutored in the philosophy of their dicipline are quick to overvalue their activity. Theorists may point to experimenters with disdain, citing them as bottle washers or meter readers; on the other hand, laboratory workers may complain about the easy life of the armchair scribbler. The disparagement of library research stems from this consideration.

No doubt some ego-building is involved in the tendency of scientists to label and reduce other science subjects as subsidiary. In some schemes, mathematical physics may rank high and experimental psychology may be a soft science. The nonquantitative, nonanalytical, and nonabstract are inevitably relegated to lower standings.

The presentation of science has varied enough from the earliest days to generate conflicting views about its communication. The semiofficial historian of the Royal Society, Thomas Sprat, was impressed by the association's "singular sobriety of debating, slowness of consenting, and moderation of dissenting." In contrast to the "yielding, compliant" temperament of the scientists, he cited the "bold and haughty asserters." Less than two hundred years later, Humphry Davy was presenting theatrical demonstrations at the Royal Institution in London, and personal attacks were published in science journals (see Chapter 6). Then, at the 137th meeting of the American Association for the Advancement of Science (Chicago, 1970) and at the 138th meeting (Philadelphia, 1971), groups of young, dissenting scientists disrupted several sessions. They used bullhorns, chanted, and called names; one indignant observer jabbed a protesting physicist with a knitting needle. The January 1, 1971, issue of the *Washington Post* editorialized: "Shouting appears to be the chief intellectual resource of the younger scientists who refer to themselves as "radicals" more or less in the manner of Nazi stormtroopers who used to call themselves "socialists." But their radicalism seems to consist of no more than a rejection of reason and an unwillingness to let anyone but themselves be heard.

Dogmatism is not the patent of the radical young among the scientists; the trait is more often evident in the established members of the community. For example, during the course of the arrangement for a symposium on unidentified flying objects at the 1969 meeting of the American Association for the Advancement of Science, several older physical scientists vigorously opposed giving credence to "unscientific" ideas, and they had enough votes to carry Section D on astronomy. They sent letters to congressmen and to the Vice-President of the United States urging invervention to cancel the symposium.

The dogmatism of established scientists is one of the factors for the minute amount of research in such realms charming the nonscientist as telepathy, clairvoyance, extrasensory perception, precognition, reincarnation, life-span extension and psychokinesis. If the public had a voice in the choice of subject matter for research they would overwhelmingly support parapsychology phenomena as well as so-called cures for cancer such as krebiozen and the apricot pit product called Laetrile. Dogmatism, too, was responsible, in part, for 5000 U.S. scientists sending a letter in November 1945 to the President of the United States claiming the inclusion of the social sciences in a proposed National Science Foundation would be a serious mistake.

Evidence, reasoning, and experiment have been the traditional arbiters of science; the facts and interpretation by peers were inviolate. In 1970, a British dental surgeon challenged this tradition by alleging that the *British Medical Journal* had libeled him. The surgeon had developed a new technique for anaesthesia and so acquired an international reputation. A committee appointed by the Ministry of Health questioned the technique's effectiveness. Four men from the Department of Anaesthesia in the University of Birmingham then tested the procedure and reported negative results. According to the *London Times* of February 14, 1970, Lord Denning, Master of the Rolls, commented, "it would be a sorry day if scientists were to be deterred from publishing their findings for fear of libel actions. So long as they refrained from personal attacks, they should be free to criticise the systems and techniques of others. Were it not otherwise, no scientific journal would be safe."

In the case of Drummond-Jackson *versus* the British Medical Association, on October 31, 1972, on the thirty-eighth day of the trial when only half of the plaintiff's case and none of the defense had been heard, an out-of-court settlement was reached. The final sentence of the agreement read in open court said that the plaintiff "recognizes and acknowledges that the *British Medical Journal* has a right and duty to its readers and to the medical profession generally to publish articles such as that submitted to them by the individual defendants and to comment on them."

The conduct of science as an international venture has also had more challenges recently. Group loyalties, particularly nationalism, have always been present while the international brotherhood of science was being developed. Niccolo Tartaglia, founder of the science of ballistics, was ashamed of the military significance of his work but became a national zealot when Italy was invaded by the Turks. John Napier, the Scottish inventor of logarithms, spent much time thinking about machines to help defeat Philip II of Spain if the latter ever invaded the British Isles.

The English have always publicized their scientific heroes extensively and often without mentioning the work or competition of others. Francis Bacon, Lord Verulam, was extolled by successive generations. Soon after Bacon died, Robert Hooke said, "No one, except the incomparable Verulam has had any concept of a plan for the guidance of the intellect in scientific research." In the nineteenth century, Sir John Herschel proclaimed, "Previous to the publication of the *Novum Organon* natural philosophy, in any legitimate and extensive sense of the word, could hardly be said to exist." Sir Oliver Lodge called him "the herald of the dawn of science," and Horace Walpole declared "Bacon was the prophet of the things which Newton revealed to man." On the other hand, nineteenth-century German chemist Justus Liebig considered Bacon to be poorly informed in science and essentially a dilettante.

Isaac Newton was deified by the English. Poets wrote couplets and odes to his honor, and luckily one of the greatest publicists of all times, Voltaire, was on the side of those who placed Newton on a high pedestal.

Michael Faraday and Charles Darwin, the two giants of nineteenth-century English science, became widely known in the non-English speaking world. Some of the fame could have been the result of Brittania ruling the waves.

During the early twentieth century, English science was spread through the journal, *Nature*, which published "The Song of the Jolly Electron" 1926 (volume 108, page 885). Anyone reading the verse would conclude that only the British were responsible for the advance of our knowledge of atomic structure. Not a single German, French, Italian, or American name is mentioned in the poem; only the Dane, Niels Bohr, is honored.

The English were as national as others. During Newton's time, Edmond Halley, after whom the comet is named, pointed out that England was an island and "must be masters of the Sea, and superior in naval force to any neighbour." The French and Germans were just as chauvinistic. Much of the work of nineteenth-century Louis Pasteur can be interpreted as his effort to help France. When France and Germany were at war in 1870, he returned his honorary medical degree from the University of Bonn and volunteered for service. Some of the Germans were equally nationalistic. In 1870, Rudolf Clausius organized a volunteer ambulance corps of students at the University of Bonn. Later, Clausius was subject to the anti-German prejudices of British physicist Peter G. Tait, who claimed priority for his countrymen in thermodynamics and spectroscopy.

In 1874, Adolphe Wurtz, professor of organic chemistry at the Sorbonne, published the first volume of his *Dictionary of Pure and Applied Chemistry* in which he said, "Chemistry is a French science." Later, during World War I, many French scientists savagely attacked German science.

The best known nationalist-scientist of the twentieth century is Fritz Haber. His nitrogen-fixation process was a great help to Germany. He was a leader in the use of poison gas during World War I and later valiantly and unsuccessfully searched for means to obtain gold from seawater so that Germany could pay reparations. However, practically all the German savants supported the war effort through word, or deed, and the same can be said for the American, Russian, English, French and Italian scientists.

Enrico Fermi is generally not considered a nationalist, yet his biographer claims that the introduction of modern physics into Italy was one of the main objectives of his early years.

The nationalism of scientists is at variance with the international character of science described at the end of Chapter 1. Internationality has been more accentuated with time and can be the dominant tendency only if scientists so desire and actively support the supranationalism of the enterprise. Scientific workers are as much subject to the pressures of belonging to the group called "the nation" as are plumbers or actresses. Whether scientists accede to this call or to the relatively nascent clarion of the world order depends primarily on the strength of their fraternity and their acceptance of the international aspect of science.

SELECTED READINGS

Burniston Brown, G. *Science, Its Method and Philosophy* (New York: W. W. Norton, 1950).

Goran, Morris. *The Story of Fritz Haber* (Norman: University of Oklahoma Press, 1967).

"Letters to the Editor." *Science* **171** (January 22, 1971), 230.

Schroeder-Gudehus, Brigitte, "Challenge to Transnational Loyalties: International Scientific Organizations after the First World War," *Science Studies,* **3** (April 1973), 93-118.

CHAPTER 5

Conflicts about Subject Matter

Controversy has been omnipresent in the development of every science, yet some scientists deny its existence. Theodore Caplow and Reece McGee in their *The Academic Marketplace* (New York: Basic Books, 1958) report interviewing many university professors who remarked, "I'm not aware of different viewpoints or schools of physics," or "There are no schools in physics," or "That's not a meaningful question in science." The arguments have been not only about ideas but also concerned with priority; in addition, today's debates deal with the impact of science and technology.

Technology was brought forcefully to the world's attention with the development of nuclear weapons. The furor about pollution made it ubiquitous and brought comparable issues to the foreground. In December 1971, the United States resolved to establish a National Advisory Commission on Health, Science and Society, a body to focus on ethical and policy questions raised by biomedical technology. Some other organizations involved in this quest include the Committee on the Life Sciences and Social Policy of the National Academy of Sciences; the Institute of Society, Ethics, and the Life Sciences at Hastings-on-Hudson, New York; and the Joseph and Rose Kennedy Institute of Bioethics at Georgetown University, Washington, D.C.

The irritating questions are not only how to care for the infirm and aged but are also about heart transplants, the detection of genetic disorders, *in vitro* fertilization of human ova, sex choice of the unborn, and cloning (producing an individual genetically identical to the original). Issues such as sterilization of the retarded, behavior modification through drugs, abortion, and denial of life-saving surgery to newborn Mongoloids having repairable physical defects are also involved.

The two American science leaders in the environmental crusade, Barry Commoner and Paul Ehrlich, have argued their position in public. One accents excess population as the culprit, and the other stresses faulty technology as the chief cause of environmental disruption. Much of the May 1972 issue of the *Bulletin of Atomic Scientists* contained their charge and countercharges.

Debate is the rule for all the topics; it is within the physical science and society area. One debate about the antiballistic missile (ABM) also brought direct charges of dirty pool.

In 1969, the issue of the ABM was argued in public and before congressional committees. Many of the vocal opponents of the weapon were based at the Massachusetts Institute of Technology; the proponents had fewer scientists, but they had Albert Wohlstetter, professor of political science at the University of Chicago. He asked the Operations Research Society of America to "appoint a panel to consider some aspects of professional conduct during the ABM debate." The six-man committee appointed November 1969, published the results of its inquiry in the Fall 1971 issue of the journal, *Operations Research*. They claimed that some ABM critics presented false or misleading arguments to congressional committees. Five of the thirteen-member Council appointing the investigating group protested the procedures followed; the society's founder, Philip M. Morse, threatened to resign. He wrote a letter to the *Boston Globe*, suggesting that the Operations Research Society of America "is on the side of ex-Senator Joseph McCarthy, is pro-military, and supports the assumption that the expert always knows best." Several members who supported the final publication and the Society responded in many ways, including letters to the editor of *Science* (published in the March 31, 1972, issue). Among others who found value in the end result was the editor of *Minerva*, a scholarly journal for science and learning. An abridged version of the inquiry's conclusions was printed in *Minerva's* January 1972 issue because "It raises, albeit with respect to a very narrow and abhorrent subject-matter, fundamental questions regarding the standards which scientists should observe in offering advice to laymen. It reminds scientists of their obligations as scientists in dealing with each other. It is not an agreeable document to read, but no publication in recent years states the facts so straightforwardly."

The principals in the ABM debate either accepted or reevaluated the data given by the Defense Department. In no instance were attempts made to censor, muzzle, or deter the participants. This freedom is not always the case.

In 1969 and 1970 at the Lawrence Radiation Laboratory operated for the Atomic Energy Commission by the University of California, debate occurred over the right of scientists to hold formal political discussions at the laboratory during their lunch hours. At the Berkeley facility, the director, Nobel laureate Edwin M. McMillan appointed a five-man committee to reexamine what he called the laboratory's traditional policy of "excluding political advocacy from the Laboratory." On February 25, the committee issued its report:

> Should we introduce in addition problems of a political and
> social nature outside of our assigned mission? We believe not.
> Moreover, we know from our sampling of Laboratory opinion
> that substantial numbers, perhaps a majority, of Laboratory

employees would resent our doing so with varying degrees of emphasis. Thus introduction of such discussions into the Laboratory proper will almost certainly lead to estrangement and division among Laboratory employees which we feel will, in the long run, impair the efficiency of our Laboratory and the solidity of its support by the community and the nation.

Another Nobel laureate at the laboratory, Owen Chamberlain, said: "I claim it should be a standard function of the laboratory staff to discuss where all forms of science are taking us, whether this involves chemical pesticides or smog or new fuel systems or nuclear weapons policies."

Early in 1973, Charles Schwartz, professor of physics at the University of California at Berkeley, won a lawsuit against the Lawrence Radiation Laboratory, more or less as a result of the ban on lunchtime politics. Because he had held two lunchtime seminars in defiance of the rule against meetings he was suspended for two weeks, later reduced to two days. In 1970, he was told that his request for summer employment at the laboratory had been denied, and he filed complaints with the proper faculty committee. They failed to find fault with the decision. Schwartz appealed without success to the Academic Freedom Committee, the president of the university, and the Board of Regents. The Berkeley branch of the American Federation of Teachers took his case to court and won. Before the termination of the lawsuit, a new policy was instituted allowing freedom of speech at the laboratory at all times.

In earlier generations, the conflict within science about the use and significance of scientific achievements was on a low-key basis. The discussions were undertaken by retired laboratory workers and the philosophically minded. Many immersed in the study of nature and technological exploits saw nuclear physicist Ernest Rutherford's statement about scientific method arguments, "all hot air," as their credo for talk about the society and science interface.

In prior generations, as now, scientists did get excited about recognition. Credit in one form or another for scientific achievement is desired by all investigators. The struggle for this recognition has at times broken friendships, scarred personalities, and left a bitter aftertaste.

Before he became well known, Galileo was active in the search for credit. In 1606, he charged that his small book about the geometrical and military compass had been plagiarized. In 1614, he accused Simon Marius of Anapach of stealing the record of his observations of the moons of Jupiter. Galileo also had a lively argument about priority with regard to sunspots.

Among the famous recognition battles of science was the seventeenth-century one between Robert Hooke and Henry Oldenbourg about the discovery of a balance spring for watches. In France in the next century, three Academicians were sent to Peru to measure the shape of the earth; they worked independently and then quarreled over their results. One of them remarked that ten years of toil in the new world were followed by ten years of disputes in the old. Paul Ehrlich, famous for finding a cure for syphilis, and Emil von Behring parted in anger about their roles in un-covering the diphtheria antitoxin for which von Behring won a Nobel prize. Chemist Frederick Soddy was in his later years bitter about lack of recognition for his work with Ernest Rutherford in radioactive disin-tegration theory.

Lawsuits are frequent when the achievement is immediately practical; patent infringement is charged. Not too long ago, Alfred Nobel sued Professor Sir Frederick Abel, a military chemist, and Professor James Dewar on this account, and a recent commentator suggests that the dividing line between their use of Nobel's work "and actually stealing the invention was very fine."

During the late nineteenth century, Christian H. F. Peters, astronomer at Hamilton College, Clinton, New York, sued his former student, Charles A. Borst, for the recovery of a star catalog. Borst had submitted a proposed title page indicating that he had performed the work under the direction of the professor, and Peters, perturbed, demanded the catalog. Although the case was resolved in favor of Peters, he never secured the manuscript. A U.S. Court of Appeals reversed the decision in 1894, four years after Peters' death. U.S. astronomers decried the process, wishing the affair had been settled by a committee of astronomers.

In our time, pure scientists have also resorted to the courts. The discoverer of streptomycin, Selman A. Waksman, was sued by a worker in his laboratory. Another outstanding scientist, Nobel prize winner Severo Ochoa, has been taken to court for allegedly preventing an associate's publication. In 1972, physicist Oreste Piccioni filed suit asking \$125,000 and an admission by two Nobel laureates, Emilio Segre and Owen Chamberlain, that the design of a 1955 experiment for which they were awarded the Nobel prize was really Piccioni's (a case still in the courts).

Now, as in prior generations, debates are about the worth of a project. The American Psychological Association is the center of one costing several million dollars. At the start of the National Information System for Psychology, a small and angry group of research psychologists claimed their interests were being ignored. Some of the plan to abstract and disseminate psychological literature was, according to the editor of the *Journal of Experimental Psychology*, a "halfwitted scheme."

Many disputes about a scientific idea are settled by argumentation, persuasion, and display of evidence by the original scientists and others. In the controversy between Luigi Galvani and Alessandro Volta about the nature of electricity, Galvani's co-worker and greatest supporter was his nephew Giovanni Aldini, a professor of physics at the University of Bologna and organizer of a society to foster Galvanism (animal electricity). The debate about whether electricity was one or two fluids was carried out be many eighteenth-century naturalists. At about the same time, the followers of James Hutton, Plutonists who believed that rocks must have consolidated from a molten condition, were embattled with those who followed Alfred Werner, Neptunists who argued that precipitation from water was the process.

During the nineteenth century, James Clerk Maxwell's mathematical presentation of electromagnetic radiation was greatly admired, but discussion about its verification prevailed until Heinrich Hertz's experiment. In the twentieth century, the phenomenon of superconductivity was not satisfactorily explained until about fifty years after its discovery, and arguments about it were plentiful.

At the beginning of the last third of the twentieth century, several scientific topics were being vigorously debated. One in the biological sciences was the cause of atherosclerosis, particularly the role of cholesterol and other lipids in the blood. In the physical sciences, controversies swirled about the merits of the claims by the U.S. and Soviet teams for the discovery of element number 104 and about the meaning of the intense flux of gravitational radiation from the center of the galaxy.

Those who have witnessed scientific debates have found them to be memorable occasions. The enthusiasm of the interested spectators is like that at a sports contest. For example, Physicist Arnold Sommerfeld recalled:

> The champion for Energetics was Helm; behind him stood Ostwald, and behind both of them the philosophy of Ernst Mach (who was not present in person). The opponent was Boltzmann, seconded by Felix Klein. The battle between Boltzmann and Ostwald was much like the duel of a bull and a supple bull fighter. However, this time the bull defeated the toreador in spite of all his agility. The arguments of Boltzmann struck through. We young mathematicians were all on Boltzmann's side.

One of the great debates in twentieth-century astronomy, in 1920 between Heber D. Curtis of Lick Observatory and Harlow Shapley of Harvard College Observatory, did not seem so later to one of the participants. Curtis was closest to current beliefs, arguing that spiral nebulae were at great distances and were star systems like the Milky Way. In 1969, Shapley wrote:

> As for the actual "debate," I must point out that I had forgotten about the whole thing long ago. . . . Then, beginning

about eight or ten years ago, it was talked about again. To have it come up suddenly as an issue, and as something historic, was a surprise, for at the time I had just taken it for granted. . . . I don't think the word "debate" was used at the time. Actually it was a sort of symposium, a paper by Curtis and a paper by me, and a rebuttal apiece.

The late eighteenth-century controversy in biology between aged Lamarck and his former student Georges Cuvier was much more poignant. The former was almost blind and the latter was ruthless in his presentation.

A controversy is not always directly apparent. The one between Professor Herman LeRoy Fairchild, glacial geologist at the University of Rochester, and Dr. O. D. von Engeln, geomorphologist at Cornell University, about glacial erosion in the Finger Lakes region of New York was hardly a debate. The men never resolved their different points of view. Sometimes poor publicity hides the argument. In 1971, the Department of Health, Education, and Welfare and the Allied Chemical Corporation found entirely different amounts of mercury in urine samples taken from workers at the company's chlorine and caustic soda plant in Moundsville, West Virginia.

Once in a while, the dispute may terminate abruptly as a one-sided affair. In 1775, the French Academy of Sciences voted to abandon reviewing any proposals dealing with the squaring of the circle, trisecting the angle, duplicating the cube, and perpetual motion. One indignant fellow who claimed to have squared the circle was once a corresponding member of the Academy. He sued, wrote letters, challenged eminent mathematicians, and pleaded before other groups, all to no avail. In the December 24, 1971 issue of *Science*, the president and past presidents of the Health Physics Society issued the following statement:

> On the third such occasion since 1968, Dr. Ernest J. Sternglass, at an annual meeting of the Health Physics Society, presented a paper in which he associates an increase in infant mortality with low levels of radiation exposure. The material contained in Dr. Sternglass' paper has also been presented publicly at other occasions in various parts of the country. His allegations, made in several forms, have in each instance been analyzed by scientists, physicians, and biostatisticians in the Federal government, in individual states that have been involved in his reports, and by qualified scientists in other countries.
>
> Without exception, these agencies and scientists have concluded that Dr. Sternglass' arguments are not substantiated by the data he presents . . . . We, the President and Past Presidents

of the Health Physics Society, do not agree with the claim of Dr. Sternglass.

For some, disputes may appear to be one sided. David Green, director of the Institute for Enzyme Research at the University of Wisconsin, told a reporter for *Scientific Research* in 1968 that the scientific establishment had ignored the theory his institute had advanced. He said, "It is amazing how difficult it is to reach the establishment. After all, I, too, have position and authority—still it is impossible. I have been fighting them for years. I could write a book about it." Books have been written about the conflict between almost the entire physics community, believers in the quantum theory and its inferences for free will, and Albert Einstein, who opted for a more strict determinism.

A conflict may be tempered or decided within the space of a generation. In the late seventeenth century, the Danzig astronomer Johannes Hevelius argued with Robert Hooke about the merits of telescopic versus naked-eye sights; Hevelius strongly advocated the latter. To settle the controversy, the Royal Society sent twenty-two-year-old Edmond Halley to Danzig for two months. Halley was satisfied that Hevelius did well with his procedure and wrote a testimonial letter offering to bear witness "to the scarce credible accuracy of his host's instruments, against all who may in future call his observations in question, having himself with his own eyes seen, not one or two, but many observations of stars made with the great brass sextant agree most accurately and almost incredibly with one another."

Another example was displayed in the January 28, 1972, issue of *Science* where George R. Price wrote: "During the past year I have had some correspondence with J. B. Rhine which has convinced me that I was highly unfair to him in what I said in an article entitled "Science and the Supernatural" published in *Science* in 1955 (August 26, p. 359). The article discussed possible fraud in extrasensory perception experiments."

William Duane of Harvard University refused for a long time to believe in the Compton effect—the change of wavelengths of X-rays when they collide with electrons. But at a meeting of the American Physical Society, he withdrew his objections, admitting that Arthur Compton's claims were correct in all respects.

Some significant scientific questions have had more than one generation of controversy. The heliocentric system, starting with the statement of Copernicus in 1543, was established only after a couple of hundred years. In 1622, Francis Bacon dismissed the idea with "In the system of Copernicus there are found many and great inconveniences." The final, persuasive piece of evidence, stellar parallax, was detected at the beginning of the nineteenth century, although intelligent people had long before adopted the idea.

Spontaneous generation has had a long history of charge and countercharge. During the seventeenth century, the Tuscan physician Francesco Redi performed one of the first experiments to show that life could arise only from other life. When he protected meat with fine muslin, maggots appeared on the cloth, not on the meat. A generation later, a Welsh Roman Catholic priest and naturalist, J. T. Needham, showed that microorganisms rose spontaneously from putrefying organic substances. Spallanzani repeated the work with more or less prolonged boiling of solutions to contradict Needham. Next, experiments by J. L. Gay-Lussac were used to support Needham's views while Theodor Schwann's work on the subject gave contradictory evidence.

The definitive work on spontaneous generation was done by Louis Pasteur. His results satisfied the French Academy of Sciences and were in line with Pasteur's religious views. Nonetheless, about one hundred years later, spontaneous generation with proper conditions was an accepted article of belief. Life began from the lifeless long ago when the earth had reducing atmosphere and the delicate sprig of existence could flourish.

SELECTED READINGS

Blissett, Marlan. *Politics in Science* (Boston: Little, Brown, 1972), 130-151.

Boffey, Philip M. "Science and Politics: Free Speech Controversy at Lawrence Laboratory," *Science,* **169** (August 21, 1970), 743-745.

Goran, M. *The Future of Science* (New York: Spartan Books, 1971), Chapter 11.

Gordon, Theodore. *Ideas in Conflict* (New York: St. Martin's Press, 1966).

CHAPTER 6

Personality Conflicts in Science

As in all human endeavors, personality conflicts abound in science. From the very start, great and small accomplishments have been laced with difficulties between people. Those prior to the twentieth century are recorded in the more critical biographies and journals of the history of science. Among the most noteworthy tiffs were the problems of Isaac Newton and Robert Hooke, wherein Newton would not accept the presidency of the Royal Society until Hooke was no longer associated with it; and of Humphry Davy and Michael Faraday, wherein Davy sponsored and aided Faraday but finally blackballed his fellowship application to the Royal Society. Lesser known personal conflicts include the name-calling indulged in between the American paleontologists Othniel C. Marsh and Edward Drinker Cope and the friendship that turned to hatred between the French naturalist Pierre de Maupertius and the writer, intellectual, and popularizer of Newton's work, Voltaire. Another soured friendship in the same country a generation later involved Jean Baptiste Biot and Dominique Arago over the wave theory of light; both men at first adhered to the particle conception until Arago became an adherent to the wave idea. All of those conflicts had bitter overtones; the one between Justus von Liebig and French chemist Dumas had comic aspects. Liebig published in the journal he edited a satire, written by his friend Woehler, on the substitution theory of Dumas and the paper was signed by *S.C.H. Windler.*

At the beginning of this century, each of the centers of scientific excellence had instances of friction between personnel. Even in England where the tenets of good sportsmanship are inculcated on school playing fields, ugly personality disputes appeared.

Ernest Rutherford from New Zealand came as a graduate scholarship student to the Cavendish Laboratory at Cambridge University. According to him, some of the advanced research students "gnashed their teeth with envy" and placed obstacles in his path. In one letter home, he described a student "on whose chest I should like to dance a Maori war-dance."

Rutherford was not angelic in all his dealings. When he was at Manchester University, he had one discussion with the chemistry department head and finally said that he was a nightmare "like the fog-end of a bad dream." Sir William Ramsay was often the target of Rutherford's barbs; at a meeting of the British Association for the Advancement of Science, Rutherford said he "got up and poked fun at his arguments." Rutherford wrote a friend about Ramsay: "He is now absolutely unreliable in everything he publishes."

In Italy, Professor Antonino LoSurdo was the antagonist of Enrico Fermi, the renowned physicist. The former tried to prevent Fermi's appointment to a prestigious position in Rome. Inadvertently or otherwise, LoSurdo misplaced an important letter nominating Fermi to the Accademia dei Lincei.

The most poignant personality disputes involve those who were once friends. In his *Science and Government,* Sir Charles Percy Snow has documented one case between Fredrick Lindemann, later Lord Cherwell, and Tizard, that developed during World War II. Nuel Davis has described another between E.O. Lawrence and J. Robert Oppenheimer in his *Lawrence and Oppenheimer.*

Current conflicts very seldom reach the stage of publication. Libel laws and the cold, impersonal tone of scientific reports prevent the transference of vituperation from the cloakroom to the printed scientific page. Newspaper and magazine articles have thus become the vehicles for the documentation of disputes, and only the extreme cases are deemed "newsworthy."

At the Research Corporation dinner, April 18, 1962, James E. Watson was honored for his work on the molecule of heredity, DNA. However, his was not a usual after-dinner speech; among other details, he recounted episodes of clashing personalities.

When the National Academy of Sciences was electing a president in 1950, a group of chemists informally organized to prevent the election of James B. Conant; there was a feeling that he had been excessively authoritarian during World War II in dealing with some of his colleagues. The conflict between Stanford Ovishinsky, founder of Energy Conversion Devices, and some members of the American Physics Society erupted into name-calling and the use of four-letter epithets. Two astronomers at Cambridge University, Sir Martin Ryle and Sir Fred Hoyle, have a personal relationship rumored "to be about a few degrees above absolute zero." In the January, 1971, issue of the British *Science Journal,* the meeting on the Social Impact of Biology, organized by the British Society for Social Responsibility in Science, was described as deserving the words "smug, elitist and glib." One Nobel prize winner was hissed by the audience when he tried to browbeat another. A personality, as well as policy, clash was revealed in the Civil Service grievance proceeding in 1971 of J. Anthony Morris, the former influenza control officer at the Division of Biologics Standards of the National Institute of Health. He claimed to have been harassed and pressured by the management of the Division and by scientists.

In a letter published in the May 10, 1968, issue of *Science*, Nobel prize winner S. E. Luria more or less summarized the issue. He wrote: "For in science, as in any other human activity, personality and competitiveness are ever-present, even determining elements . . . . How long will students who read of the wonderful epic of the "fruitfly room" at Columbia not be told of the jealousies and tensions that were part and parcel of its scientific history?" About four years later in the February 4, 1972, issue of the same journal, Michael T. Ghiselin of the Department of Zoology, University of California, reviewing *The Origins of Theoretical Population Genetics* by William B. Provine (University of Chicago Press, 1971), wrote: " . . . he notes that personality conflicts are sometimes very important in the development of scientific ideas. I dare say that this should be given as a rule rather than an exception, and the rule casts a great deal of light on the history of biology."

The last frequent personality conflicts are those occurring within the scientist himself—the kind that produce mental illness or suicide. A few scientific workers have been temporarily unbalanced, generally as a result of overwork or lack of recognition. Among the famous who suffered nervous breakdowns are Isaac Newton, Marie Curie, James Clerk-Maxwell, and Michael Faraday. Those who spent some time in an insane asylum include Robert Mayer and Ignatz Semmelweis, and both were probably hospitalized for the same reason—lack of acclaim by their peers. Suicides are relatively rare. When in a depressed state, physical scientists Max von Pettenkofer, Ludwig Boltzmann, Bertram B. Boltwood, Wallace Carothers, Paul Ehrenfest, and Percy W. Bridgman took their own lives. Biologist Paul Kammerer died by his own hand after his experiments dealing with the inheritance of acquired characteristics were reported to be fraudulent; unrequited love might also have been a cause in his case. Biochemist Hans Fischer committed suicide after the air raids by the Allies on Munich during World War II destroyed his laboratory, and biochemist Rudolf Schoenheimer destroyed himself when the Nazis appeared to be winning. Geophysicist Everette Lee De Golyer committed suicide after being ill with aplastic anemia for seven years. Chemist Emil Fischer died of a drug overdose. Chemist A. Gutbier, head of the chemistry department of Jena University, early in the present century took his own life, as did chemist Otto Hönigschmid of the University of Munich after his home was requisitioned by United States troops in 1945. If physicians are to be included as scientists, then many more names must be added to the list of those who have totally mutilated their bodies and minds.

SELECTED READINGS

Clark, Ronald W.   *Tizard* (Cambridge: MIT Press, 1965).

Davis, Nuel Parr.   *Lawrence and Oppenheimer* (New York: Simon and Schuster, 1968).

Reif, F.   "The Competitive World of the Pure Scientist," *Science,* **134** (December 15, 1961), 1957-1962.

Snow, Sir Charles Percy.   *Science and Government* (Cambridge: Harvard University Press, 1961).

PART TWO

Myths

Well-known personalities, historically significant or not, as well as topics affecting many people are prone to the development and dissemination of myths. In American history, for example, the Parson Weems tale about George Washington confessing to cherry tree destruction is widely believed. Science and scientists also have apocryphal stories: Galileo is supposed to have dropped weights from the Leaning Tower of Pisa, and Isaac Newton reportedly came to his gravitation idea after noticing an apple fall from a tree. Sometimes such accounts can be checked for truth; for example, Thomas Edison denied he had ever been beaten on the ears by a train conductor with blows so brutal as to bring on deafness later in life.

In some circles, a widely distributed and believed myth is that scientists are hermitlike seekers after the truth, occasionally badgered by non-scientists; the accounts presented in Part I belittle such an idea. In Part II, the first two chapters deal with more myths about scientists, and the last two chapters concentrate more on the subject than on the individual.

CHAPTER 7

Myths about Scientists

The image of the scientist as being distant from the arts is not consistent with the facts. Many scientists have been and are poets, writers of fiction, musicians, and fine artists. Jacobus Henricus van't Hoff, the first Nobel laureate in chemistry, seriously considered becoming a poet. In his lecture, "Imagination in Science," October 11, 1878, he named fifty-two people, from a list of two hundred famous scientists chosen at random, who showed presence of an artistic inclination, generally toward poetry. He failed to mention many others, some of whom are described below.

In his medical poem published in 1530, Girolamo Fracastoro was the first to use the term *syphilis*. A little later, the pioneer military surgeon Ambroise Paré also produced verses, but he may have borrowed them from Jean Lyége, author of Latin verses on human physiology. Ichthyologist Salviani wrote a somewhat pornographic composition in 1554. John Baptista Porta, an organizer of one of the first scientific societies, wrote many comedies and tragedies.

During the seventeenth century, Francesco Redi, known for his work on spontaneous generation, produced poetry. Even the genius Isaac Newton composed a few lines of verse.

Italians, Englishmen, and Russians are among the eighteenth-century scientist-poets. Alessandra Volta was a physicist, Mark Akenside and John Armstrong were English physicians, and Mikhail Lomonossov was a Russian universalist, but all wrote poetry.

Humphry Davy, an outstanding chemist during the last century, was friends with poets Samuel Taylor Coleridge and William Wordsworth. Wordsworth once said that the only two people to whom he felt inferior as a poet were Coleridge and the Irish mathematical physicist William Rowan Hamilton, thirty-five years younger than he. Between the ages of seventeen and twenty-four, Hamilton could not decide whether to abandon mathematics for poetry. Later, he said, "I live by mathematics but I am a poet." He and astronomer John Herschel exchanged sonnets. Scientist-inventor-physician Erasmus Darwin, grandfather of Charles Darwin, published his long poem *The Botanic Garden* in the early 1790's. Coleridge called him "the first literary character in Europe." James Clerk Maxwell and William Thomson, later Lord Kelvin, were other British physicists of the time who wrote poetry. *Nature* for May 30, 1872 contained sixteen lines entitled "Electric Valentine," published anonymously but written by James Clerk Maxwell. During the nineteenth century, physicist William Rankine at one time wrote a song in defense of the English system of units. John Keats was a famous poet, but in his short life span of twenty-five years, he had been apprenticed for five years to a general practitioner and had registered for medical school.

American professionals can also be included. The founder of the Society of Surgeon Dentists of New York, Eleazar Parmly, and the first editor of the first American dental periodical, S. Brown, were interested in poetry. Brown wrote a poem on the diseases of the teeth and their proper remedies.

Nobel prize winning biochemist Otto Meyerhof wrote poetry until the year of his death, 1950. Poems continue to appear in scientific journals. The November 1948 issue of *Physics Today* contained "Take away Your Billion Dollars" by Arthur Roberts.

The January 15, 1971, issue of *The Journal of Organic Chemistry* printed a verse entitled "Comparative Mobility of Halogens in Reactions of Dihalobenzenes with Potassium Amide in Ammonia." *Industrial Research* (April 1971) commented with four stanzas describing the publication. The respected English journal *Nature* published a geological scientific report in verse in the July 2, 1971, issue. In 1972 *Bioscience* published an epic poem, composed by a premedical student at the University of Illinois, on the origin of the eukaryotic cell. At the 26th annual conference on Engineering in Medicine and Biology, Minneapolis, physician Howard Shapiro presented his paper "Continuous Redox Monitoring of Preserved Organs" as "Talking Redox Blues" in poem form, accompanying himself on the guitar while delivering the paper.

According to the poet Saint-John Perse, Albert Einstein said to him that the poet and the scientist are akin in having a sudden illumination. Discovery and invention were logical, intellectual, and analytical after the great forward leap of the imagination.

Today's scientists are also writers of fiction, generally science fiction. John R. Pierce, an expert in electronics, is the author of eleven such books under the pseudonym of J. J. Coupling. Biochemist Isaac Asimov has made a career of writing science fiction as well as popular science. Other writer-scientists include British physiologist W. Grey Walter, American psychologist B. F. Skinner, British astronomer Fred Hoyle, British-American scientist Jacob Bronowski, American chemist Harrison Brown, and French biologist René Dubos. Charles Percy Snow gave up physics and government service for novel writing.

Biochemist Ernest Borek, in a popular book published in 1965, compared the realm of science to that of art. He wrote: "I recall that when I first read Dr. Vincent du Vigneaud's brief description of his synthesis of the hormone of the posterior pituitary, I experienced the same spine-tingling excitement as when I first heard Toscanini conduct Beethoven's Ninth Symphony, or when I first read Melville's description of Pequod sailing through the night, sails billowing and the fires under the melting kettles sparking the dark Pacific night."

Lawrence Durrell wrote: "Science is the poetry of the intellect. Organic chemistry contains some of the noblest—and least appreciated—passages in the collected volumes of the poetry of the human mind."

Some scientists, for example, nineteenth-century Sir Charles Bell, who drew all the illustrations in his anatomy books, have been fine artists. German zoologist Ernst Haeckel at one time seriously considered painting as a profession; he made hundreds of compositions anyway. At the age of eighteen, Louis Pasteur received a certificate in arts. He made many attractive pastels and lithographs and drew anyone who desired a portrait. American chemists have long had exhibitions of their paintings during their national meetings; the fourth such showing of art by chemists occurred at Cleveland, November 27-30, 1956. The tenth international exhibition of art by chemists was held in Chicago, August 1973, during an American Chemical Society meeting.

An exhibit entitled "Art in Science" brought many compliments. Art critic Robert Coates wrote in the *New Yorker* (November 16, 1965): " . . . it is frequently impossible to determine whether a given item . . . is a painting or a sculpture by one of our modern masters or a more or less run-of-the-mill production (a laboratory photomicrograph, say, of a fragment of fish roe) that just happens to look like an abstract Expressionist painting." Biologist C. H. Waddington wrote *Behind Appearance* (M.I.T. Press, 1970), a study of the relations between painting and the natural sciences in this century.

In the December 20, 1971, issue of *Chemical and Engineering News*, the editors wrote: "We at C and EN have often wondered what chemists do away from their desks or lab benches. The image of them always buried in textbooks or conducting Boris Karlofflike experiments in maniacal seclusion is just too absurd." They commissioned a writer to make a quick survey of chemist-musicians. He found many, including Jerome Hines, who became a professional singer, and Dr. DeRocco of the University of Maryland, who said "I would die without music." The most famous chemist-musician was the nineteenth-century Russian Alexander Borodin, composer of the well-known opera, *Prince Igor*. Married to a brilliant pianist, he was also a leader in founding a medical school for women. Currently famous are Martin Kamen, a violist and the codiscoverer of carbon-14, and Waldo Cohen, a cellist and originator of the ion exchange separation of nucleic acids.

Twentieth-century mathematical physicist Warren Weaver summed the artistic inclinations of scientists with: "Science is an essentially artistic enterprise, stimulated largely by curiosity served largely by disciplined imagination, and based largely on faith in the reasonableness, order and beauty of the universe of which man is a part."

   Since many scientists are involved with the arts in one way or another, one of the myths surrounding scientists can be laid to rest; that is the popular view that laboratory people see science as the be-call and end-all of existence and as the panacea for all the troubles of mankind is false. Scientism, considering the engagement of scientists with the arts, does not seem to stem from scientific workers. Even those scientists who are not associated with traditional arts have spoken out against scientism. A dean of twentieth-century American engineering, Vannevar Bush said; "Much is spoken today about the power of science and rightly. It is awesome. But little is said about the inherent limitations of science and both sides of the coin need equal scrutiny." On another occasion, he reported: "Science, by itself, provides no panacea for individual, social and economic ills. It can be effective in the national welfare only as a member of a team, whether the conditions be peace or war. But without scientific progress, no amount of achievement in other directions can insure our health, prosperity and security as a nation in the modern world." While he was president of the National Academy of Sciences, Detlev W. Bronk commented: "The thing that disturbs me is the idea that science can solve everything. For instance, when it is pointed out that we are using up a natural resource at a terrific rate, even relatively thoughtful people will say, 'yes, but why worry— science will find an adequate substitute.'" Mathematical physicist and science administrator Warren Weaver confessed: "As dedicated as I am to science, I am definitely one of the individuals who do not feel that science is the answer to all of life's problems." In 1965, the *Bulletin of Atomic Scientists* published the following confession by Nobel laureate in physics Max Born: "Though I love science I have the feeling that it is so much against history and tradition that it cannot be absorbed by our civilization. The political and military horrors and the complete breakdown of ethics which I have witnessed during my life may be not a symptom of an ephemeral social weakness but a necessary consequence of the rise of science—which in itself is one of the highest intellectual achievements of man."

   Some contend that the scientist and the artist are akin in being oblivious to the daily stress and tension to which the average person is subject. The artist and scientist are also classified as impractical dreamers. Both professionals, of course, need imagination; but the scientist, at least, is far from aloof to worldly concerns.

   Many scientists have amassed wealth, giving lie to the myth that all scientists are poor and struggling. Isaac Newton began as a lad from the farm and died a rich man. William Thomson amassed a fortune through his patents. In this century, Rudolf Diesel was a millionaire before his fortieth birthday. The size of J. J. Thomson's estate at his death amazed his contemporaries. Thomas Edison became a wealthy man; so did physical chemists Walther Nernst and Fritz Haber and American geologist Charles K. Leith. In 1964, six American scientists, including three Nobel laureates, formed Quadri-Science, to provide, for a fee, scientific advice for industry. Others have started small manufacturing companies that have grown to larger size. One unusual physical scientist told a reporter from *Science* magazine that, during the years 1959-1964, he averaged $50,000 a year in consulting fees. The contemporary American polymath, Richard Bellman, holder of three professorships in mathematics, medicine and engineering at the University of Southern California, recently read about the affluent professoriat. He said: "There's this professor here who says he's getting $30,000 a year for just thinking. For just *thinking*, he's underpaid."

Money spells power, but scientists have very little power in any temple of modern life. The myth is that in our age of science, the men and women of the laboratory exercise some degree of control. But the centers of power in education are not in the laboratory; the number of science and engineering-trained who are part of school administrations is tiny. The molders of curricula are most often the more articulate nonscientists. Very few scientists have influence in the legislative and executive chambers of government. An article in the September 24, 1971, issue of *Science* claims that the scientific advisory system in force since World War II, wherein scientists and engineers go to Washington to give counsel, has little effect on the broad technical decisions made. Two physicists at Stanford University, Frank von Hippel and Joel Primack, studied seven cases of technology and science intertwined with political decisions and concluded that science advisors are not valued for their opinions but for the appearance of legitimacy that they give to political behavior. In their words, "An advisor often finds himself playing the role of high priest whose ministrations during the preparation of a policy are supposed to make the policy immune from political attack."

The situation is similar elsewhere. England's Prime Minister Winston Churchill had a talk about the atomic bomb with the eminent physicist Niels Bohr and retained a very disagreeable memory of the interview. However, Churchill did listen often to physicist Frederick Lindemann, later Lord Cherwell.

After the decline of government funds for research during the late 1960's, scientists were advised to take their case to the public and, in some instances, to become politicians. The record of scientists in public life is dismal.

Other than those cited in Chapter 1, very few scientists have served governments. During the nineteenth century, French chemist Berthollet held some government posts, and Italian chemist Cannizarro was a functionary. Physicist Francois Arago was the navy and war minister in the provisional government of the second French Republic after the 1848 revolution. Chemist Chaim Weizmann was the first president of Israel; and chemist William Katchalsky its fourth. In the U.S. Congress, laden with attorneys, a scientist has only occasionally strayed into the chambers; less than half a dozen scientists can be counted between the years 1900 and 1971. For example, a man with a B. S. in physics from the University of California at Los Angeles served 1963-1971. The same situation occurs in England where in the early 1970s, a young animal geneticist was in the House of Commons. In the United States, a PhD chemist was elected governor of Delaware in 1968, but this was an oddity rather than a trend. Likewise, in 1973, Senator Dewey F. Bartlett of Oklahoma was a geologist (and former governor of the state), Congressman Mike McCormack of Washington was a chemist, and Congressman James D. Martin of North Carolina was an associate professor of chemistry at Davidson College.

The sparsity of scientist-politicians is paradoxical in the light of the intensity of political machinations and intrigue within the laboratory; as in every other human activity, politics appears with the presence of two or more people. President Woodrow Wilson said as much for the college campus when he confessed learning his politics at Princeton University.

During the early 1960's, social scientist Renato Taguiri polled religious figures, research managers, executives, and scientists about their value systems. He found that all ranked truth high, but scientists placed political values (such as power) higher than did the others.

The myth is that scientists are immune to the formation of groups seeking similar aims, campaigning, proselyting, caucusing, and indulging in the less-honorable activities of politics. The facts are otherwise. In June, 1973, the president of the American Chemical Society organized the Committee of Scientific Society Presidents. Thirteen organizations responded "to work for the development of a constructive national science policy." Early in 1974, the Committee of Scientific Society Presidents met with Vice President Gerald R. Ford and discussed such topics as energy, the federal management structure for science, and professional employment. The British Association for the Advancement of Science began in 1831 because the Royal Society was split by the issue of whether professionals or amateurs with a peripheral interest in science should be in control. America's pioneer oceanographer and meteorologist, Matthew F. Maury, was beleagured politically many times. According to his biographer, "the plan for a national weather bureau might have gone through but for the opposition of the scientific bloc led by Joseph Henry and Alexander Dallas Bache."

Politics within twentieth-century science has been described in asides in Nobel laureate James D. Watson's *The Double Helix* (New York: Atheneum, 1968). More extended treatments are found in the news items about science in *Nature, New Scientist,* and *Science.* Only politics can explain why two American Nobel laureates, Philip S. Hench and Frederick C. Robbins, were never elected to the National Academy of Sciences. Politics must be a factor in the assigning and acquisition of funds as well as the initial financing of projects. About half of the academic natural scientists and engineers in the United States surveyed in 1969 under the sponsorship of the Carnegie Commission on Higher Education charged that the most successful men in their fields gained their positions more as "operators" than as achievers in science.

Political adroitness is usally found in the entrepreneurs and promoters of science. One of the earliest was Baron Friedrich Wilhelm Alexander von Humboldt. Not only did he encourage young men who later became outstanding in scientific activities, he also organized, in 1827, one of the first national scientific conferences. In our time, talent is found in many places. The August 30, 1968, issue of *Science* described Wendell A. Mordy, director of the Desert Research Institute of the University of Nevada, as "an aggressive entrepreneur of science." When *Scientific Research* (October 28, 1968) wrote about Philip Handler, due to become president of the National Academy of Sciences, they called him "a man of rare ability and great incisiveness who knows his way around both Washington and the scientific community." He was in the tradition of the renowned earlier fundraisers for science, Benjamin Thompson, Count Rumford, and George Ellery Hale (the American astronomer whose talents brought the two largest observatories in the United States to fruition). Dr. Sol Tax, professor of anthropology at the University of Chicago, is another energetic American entrepreneur of science. He was most responsible for the Ninth International Congress of Anthropological and Ethnological Sciences, Chicago, September 1973. To finance the Congress, he persuaded a Dutch publisher to advance a quarter of a million dollars in prepublication royalties. He organized a volunteer work corps of about 300 people, mostly anthropology students.

SELECTED READINGS

Ashmore, Jerome. "Some Reflections on Science and the Humanities," *Physics Today,* **16** (November 1963), 46ff.

Bulletin of Atomic Scientists. February, 1959.

Cantor, S. M. "In Art: A Lesson for Chemistry," *Chemical and Engineering News* (November 5, 1956), 5406ff.

Goran, M.   "Science as Art," *Journal of General Education,* **18** (January 1967), 281-288.

————."Scientists Really Like Money," *Engineering Opportunities,* **2** (November 1964), 14, 28; also *The Chemist* (February 1966).

van't Hoff, J. H.   *Imagination in Science* (New York: Springer-Verlag, 1967).

Perl, Martin L.   "The Scientific Advisory System: Some Observations," *Science,* **173** (September 24, 1971), 1211-1215.

Stricklad, Donald A.   *Scientists in Politics, The Atomic Scientists Movement, 1945-46* (LaFayette, Ind.: Purdue University Studies, 1968).

CHAPTER 8

More Myths about Scientists

A persistent myth at any school is that scientists are brilliant. This belief is widespread among students majoring in nonscience areas, and even infiltrates the laboratories. More sober analysis yields a different estimate.

An enlightened scholar, Jacques Barzun, in his *Science: The Glorious Entertainment* (New York: Harper and Row, 1964), claims that men and women of average ability make adequate scientists. He wrote: ". . . science is the democratic technique par excellence. It calls for virtues which can be learned—patience, thoroughness, accuracy . . . Hence with care and industry a man of normal endowments can be a satisfactory scientist."

Serologist Reuben L. Kahn, in his article, "The Inspiration of Research" (*Michigan Quarterly Review,* 1965), substantiated the theme with: "Students have frequently told me that they were brilliant enough to turn to research, but I think that they most likely lacked not the brilliance but the interest and the dedication. I, for one, was far from a brilliant student at college. There were a number of far more brilliant students than I who accomplished relatively little in their careers."

It is well known that Albert Einstein was the equivalent of a high-school dropout and that Max Planck found economics too difficult. At the start of the century, enough scientists had done so poorly with science when they were in school that physical chemist Wilhelm Ostwald could say that those who are poor students are destined to be discoverers. James D. Watson, in his *The Double Helix* (New York: Atheneum, 1968), calls some scientists "stupid."

If scientists were brilliant, they would have no difficulty with understanding new theories. Yet the genius in experimental physics, Ernest Rutherford, claimed, half in jest, that the Anglo-Saxon mind could not comprehend the theory of relativity. In his 1911 presidential address to the American Physical Society, William Francis Magie said: "I cannot see in the principle of relativity the ultimate solution of the problem of the universe. A solution to be really serviceable must be intelligible to everybody, to the common man as well as to the trained scholar. All previous physical theories have been thus intelligible."

If scientists were brilliant, mistakes would be less frequent. The progress of any science can be interpreted as a succession of correcting prior errors; even a view of individual achievements shows plenty of mistakes. In veritably every science, ideas have been formulated and quickly cast aside. Tycho Brahe's planetary system was discussed during his lifetime but forgotten as soon as he died. The mistaken concepts of less prestigious scientists have had shorter life spans. At the middle of the nineteenth-century, Karl von Reichenbach, the discoverer of creosote and paraffin, proclaimed two kinds of force in a magnet; one attracted iron, and the other acted on the nervous system. He postulated the latter to be od, a physical force intermediate between electricity, magnetism, heat, and light and responsible for hypnotism. In chemistry, Canadian metallurgist A. G. French announced in 1911 his discovery of a new metal which he named canadium. The material is now known as one of the many spurious platinum-type elements claimed to have been discovered. In astronomy, Schiaparelli painstakingly observed Mercury to measure its rotation period. His value of eighty-eight days was confirmed repeatedly, yet in 1965 the rotation period of the planet was accepted to be fifty-nine and a half days. Likewise, a moon of Venus was supposedly detected in 1645, and its existence was substantiated many times; but Venus has no moon.

During the early twentieth century, French scientists wrote at length about "N rays," and more than a hundred articles were published before the idea was dropped. Later, meetings of physicists heard about the subelectron and the flow of magnetism. During the last third of the twentieth century, earth scientists have the company of a man who says that the outer one thousand kilometers of the earth is debris from a disrupted planetary body.

The many unacceptable ideas in science should be presented along with those finally succeeding. Students and others will then be able to understand that not all rejected concepts are destined for survival and later acclaim.

Many earlier as well as current scientists admitted to mistakes. Astronomer Johannes Kepler said: "How many detours I had to make, along how many walls I had to grope in the darkness of my ignorance until I found the door which lets in the light of truth." Physicist Michael Faraday claimed: "The world little knows how many of the thoughts and theories which passed through the minds of the scientific investigator have been crushed in silence and secrecy by his own severe criticism and adverse examination; that in the most successful instances, not a tenth of the suggestions, the hopes, the wishes, or the preliminary conclusions have been realized." In the second half of the twentieth century, chemist James B. Conant wrote: "One could write a large volume on the erroneous experimental findings in physics, chemistry and biochemistry which have found their way to print in the last hundred years; and another whole volume would be required to record the abortive ideas, self-contradictory theories and generalizations recorded in the same period." Physicist Francis Bitter confessed:

It turned out that once more in my life I had made a bad mistake of calculation—it was a real mistake in understanding. The layman may have a peculiar idea that all scientists worthy of the name know what they are doing perfectly. But life is too short for us to know all the aspects of everything we do. We must be satisfied with thorough knowledge in some parts, and then a certain amount of guesswork and help from others in surrounding areas. And many of us, maybe not the most distinguished scientists but certainly many competent ones, do make honest mistakes that have to be corrected by their colleagues.

Albert Einstein once wrote that "death alone can save one from making blunders." Yet many scientists are not bright enough to recognize that they can make mistakes. The early twentieth-century French chemist Henri LeChatelier wrote in one of his books: "Men who are capable of modifying their first beliefs are very rare." Also Nobel Prize winner John C. Eccles wrote in 1973: "Scientists are often loath to admit the falsification of their hypotheses, and their lives may be wasted in defending what is no longer defensible." However, enough examples of error admission by scientists are available to serve as fodder for the educators wishing to show how scientists admit their mistakes. Chemist Fritz Haber once acclaimed the conversion of base metals into gold, but within a short time, he had to revise his opinion and denounce the effort. In 1969, Soviet scientist Boris V. Derjaguin announced his discovery of an odd, new substance said to be a version of water; in 1973, he publicly changed his mind.

If scientists were brilliant, they would not have occasional lapses when they are away from their subject. Physicist Enrico Fermi's physicist-biographer Emilio Segré recounts how Fermi took up fishing: "and he developed theories about the way fish should behave. When these were not substantiated by experiment, he showed an obstinacy that would have been ruinous in science."

The supposed brilliance of scientists is at times described as the talent which allows them to cut through nonessentials to get to the heart of the matter. Pertinent facts and logic alone is said to be what sways them. Yet almost any major idea had great difficulty in being accepted. Max Planck complains bitterly in his scientific autobiography that not once in his career did he ever have the satisfaction of convincing everyone he was correct. His famous comment was to claim how ideas in science change not because of evidence but only because the old guard dies and a new generation appears. A letter from a University of Minnesota astronomer, William J. Luyten, published in *Science* (July 17, 1964), states: ". . . only thirty years ago H. N. Russell, then the dean of American astronomers, propounded the thesis that our solar system must have had a well-nigh unique origin—and the vast majority of astronomers followed him. In fact, during the thirties I was virtually the only astronomer who dared criticize the collision theory—and I was very nearly 'read out of the party' for that offense." The same resistance can be recounted for industrial processes and products. The electrostatic copying machine, for example, had a lengthy battle for acceptance.

The brilliant-scientist liturgy may be transposed into recounting the difficulty of the subject matter of science. Here the facts are contrary because nonscience topics have a larger number of factors; variables are few within the laboratory.

The intent of citing difficulty may have been to stress the abstract aspect of the subject. The concepts are more removed from direct sense impression and the human element seems lacking. This characteristic of most of science may be the reason for the highly vaunted objectivity and the detachment, disinterest and dispassion associated with it.

Objectivity in science is a myth but not in the sense of Roszak and Buckley (pages 25-26). The men and women who work in the laboratory cannot divorce themselves from their interests and feelings any more than can attorneys and theologians. The last two named have built attachments and loyalties to persons, groups, and ideas; so have the scientists. Both scientist and nonscientist are subject to the primeval urges of self-preservation, the culturally conditioned callings of the ego, and whatever lies below our conscious level and drives us. The best that can be said for the scientist being objective is that he deals with abstractions and is more accustomed to being oblivious to human beings. Then, too, some admit having a subjective approach in their work. Among the scientists who have confessed to being drawn to the mystical and metaphysical are physicists Erwin Schrödinger and Max Born.

The National Academy of Sciences in the United States began to recognize the lack of objectivity when it considered a policy in 1972 to have the approximately seventy-five hundred scientists, engineers, and their specialists serving on more than four hundred academy advisory panels, file statements attempting to outline prospective biases. An organization apparently at the other end of the establishment spectrum, the Center for Science in the Public Interest, set up by four alumni of consumer-crusader Ralph Nader's group, have a similar view. They report "the myth of objectivity is the worst myth we've got in the scientific profession."

The objective attitude can be traced to seventeenth-century Francis Bacon, a pioneer publicist for science. He had the personality to take part in what he called the "disinterested observation of nature." A twentieth-century critic described him: "If there were a thermometer to measure the intrinsic force of human passions, we should find, in the case of Bacon, that the degree of warmth belonging to his heart stood very close to zero."

Many chroniclers of science begin their accounts with Bacon or Galileo, and this fact may be responsible for the myth that science arose with them. Such an erroneous point of view ignores the achievements of earlier men and women as well as the named and unnamed geniuses of the classical world. The works of Archimedes, the greatest physicist and mathematician of antiquity, rival those of Isaac Newton or Albert Einstein. Aristotle made such contributions to biology that Charles Darwin was prompted to remark he learned much from the Greek philosopher.

No doubt, our biological heritage should also be credited. Rough equivalents of the elements of the scientific procedure can be cited in other organisms. Varieties of hypothesis-making and verification can be delineated in many animals, and organic evolution may be the history of science before man.

Our achievements in probing nature can be compared to a staircase having a variety of grades and several landings. The fitful progress prior to the Greeks is represented by an ascending section of stairs with many level places symbolic of the times when less progressive peoples were dominant. For some interpreters of man's record, a descending staircase should be invoked at times. The Golden Age of Greece would be an almost vertical staircase to show the spurt of activity. To portray the long sway of the Romans and the early Christians, the stairs would not rise at all. The sequence from Galileo through Newton would again be an almost vertical ascent. To show the almost frenetic activity from World War II to the moon landing in 1969, not only would the staircase be vertical, but the rate of ascent by man would be accelerated.

Another myth that permeates the story of scientific accomplishment through the ages is that the achievers who periodically appear are lone geniuses without sponsorship. The truth is that veritably every notable scientist was first promoted into the world of learning or else was a forgotten genius resurrected by later generations.

Galileo was unemployed and thought of moving to the Byzantine empire before his sponsor placed him. Galileo approached the man who first helped him, Guido Ubaldo, by sending him, for comment, some theorems in theoretical mechanics. Thereafter, Galileo was aggressive and talented enough to make his own way.

Friendly colleagues are often sponsors. At the end of the eighteenth century, French chemist Antoine Fourcroy was the patron of Louis Vauquelin, the discoverer of the element chromium. Vauquelin then supported Fourcroy's maiden sisters as well as sponsoring Louis Thenard, discoverer of hydrogen peroxide.

The most famous case of a scientist helping a junior colleague or a young hopeful is that of Humphry Davy's giving Michael Faraday a laboratory job. A little earlier in France, biologist Jean Lamarck was a teacher and sponsor of Georges Cuvier. In both cases, the relationships soured at the end. More often, the recipient of a favor is grateful. In the preface to his first book, on meteorology, atomic theorist John Dalton paid homage to one of his first teachers in science.

Having a sponsor is not a phenomenon of the past. In Italy, Enrico Fermi had the helping hand of Senator Orso Corbino, also a physicist. The situation in Germany is indicated by the words of the English translator of chemist Richard Willstätter's memoirs: "The climb up the academic ladder was normally slow and uncertain and rarely attempted by young men without independent means and useful connections." In the United States, sociologist of science Robert K. Merton found that thirty-four of fifty-five American Nobel laureates in science had worked in some capacity, as young men, under other Nobel prize winners.

The sponsorship need not be an extensive involvement; it could be a simple push. When amateur radio astronomer Grote Reber sent his paper containing the first radio map of the heavens to the *Astrophysical Journal,* the scientific advisors told the editor not to publish it. Otto Struve, however, did print the material.

During the nineteenth century, Baron von Humboldt innumerable times obtained financial support for mathematician Ferdinand Eisenstein. Without such sponsorship, some of the work during the twenty-nine years of Eisenstein's life may never have occurred.

Sponsorship is not necessarily an altruistic effort in all cases. The contemporary Russian physicist P. L. Kapitsa confessed: "As you grow older, only your young students can save you from a premature hardening of the brain. Each student working in his field is also a teacher. Who else teaches the teacher but his own pupils?" Kapitsa quoted his own teacher, Ernest Rutherford, as saying: "You know that only thanks to my students do I keep on feeling young."

When a young scientist is sponsored, his work at first is not necessarily in the laboratory or at the highest levels of operation. A friend of Humphry Davy's, noting Michael Faraday's plea for employment, advised giving the boy bottle-washing chores.

A myth as pervasive as the one claiming that all scientists are brilliant is the one identifying all scientific activity with the laboratory. Modern scientists do a large assortment of work. Administrator, writer, and teacher are some jobs qualifying for the title; a mathematical physicist may sit with pencil and paper and have the title. Remember, Galileo and Newton spent a large proportion of their time writing.

SELECTED READINGS

Goran, M. "Scientists Also Make Mistakes," *Chemistry,* **37** (April 1964), 28-29.

Holton, Gerald. "The False Image of Science," in Louise Young (Ed.) *The Mystery of Matter* (New York: Oxford University Press, 1965), 637-648.

Storer, Norman W. *The Social System of Science* (New York: Holt, Rinehart & Winston, 1966).

Watson, James D. *The Double Helix* (New York: Atheneum, 1968).

CHAPTER 9

Pure and Applied Science

Physicist Louis Ridenour wrote, in the August 1947 issue of the *Bulletin of Atomic Scientists:*

> I have a friend who is a band spectroscopist on the faculty of a large state university. He has been particularly interested in the band spectrum of the element nitrogen. He once said to me: "When the representatives of the state legislature visit me, I always tell them I am trying to make better fertilizer." There is to be sure, nitrogen in fertilizer and knowledge is power. It is just conceivable that my friend's investigation of the band spectrum of nitrogen may some day affect the fertilizer industry in some unexpected way. But it is undeniable that his interest is in spectroscopy, in and of itself.

An identical situation exists in the biological sciences wherein scientists may claim that cancer could be conquered through their really unrelated efforts. During the last part of the twentieth century, ecology and environment became the key words in the subterfuge.

So-called pure scientists have not always tried to rationalize, or lie, about their activity in terms of practicality. Only the increasing acceptance of the myth of separation between pure and applied science has stepped up the attempts at justification. From the beginning of modern science until recently, the distinction between science and its application was dim, if it was at all recognized. The pioneers were active in all phases of the subject.

Thales of Miletus, mentioned as impractical in Chapter 3, was a practical man when he cornered the olive press market. Leonardo da Vinci is well known for his designing of mechanical devices as well as for his efforts in pure theory. It is reasonable to assume that Copernicus was motivated by the concerns of his Church for an efficient calendar. He may have been the first to introduce buttered bread when he provided food for the men defending his castle against the Teutonic knights. Historian Lynn White, Jr., wrote in *Galileo Reappraised,* issued by the University of California Press in 1966: "During Galileo's early career—down to his acquisition of the telescope in 1609 at least—his environment and interests were largely technological." Even as a famous man, Galileo was interested in the practical. Just before he died, he submitted an unsatisfactory method for finding longitude at sea, competing for the prize offered by the

States-General of the Netherlands. Blaise Pascal, known to science students for his principle about fluids, built the first calculating device. The great German contemporary of Newton, Gottfried Leibniz, said: "The value and even the mark of true science consists, in my opinion, in the useful inventions which can be derived from it."

At the start of the nineteenth century, the Royal Institution of England was established and somehow achieved the image of a pure science establishment. However, the managers were largely landowners interested in applying science to agriculture and agricultural products. A set of instructions to Humphry Davy in 1801 is illuminating. He was told "to absent himself from the Institution during the months of July, August, and September, for the purpose of making himself more intimately acquainted with the practical part of the business of Tanning."

Many of the twentieth-century greats of pure science had an abiding interest in applied science and, at times, accomplishments in the realm. Early in his career, the father of nucleonics, Ernest Rutherford, designed a device to measure vibrations; he told his future wife, "I hope to get a fee of at least fifty pounds." Fritz Haber, the developer of nitrogen fixation, had many achievements in technology. Even that paragon of the abstract and impractical, Albert Einstein, once thought about patenting a refrigeration device.

A tentative separation between pure and applied science began to appear with specialization and division of labor. Naturalists could no longer undertake all kinds of investigation and had to be content with one branch of the subject. Philosopher Aristotle could write about biology and physics; nineteenth-century philosopher and scientist Ernst Mach wrote only about physics, particularly mechanics. The division of labor in the twentieth century brought more refinement, so that the scientist in a field as broad as physics had to be in solid state, weak interactions, or any of a number of other specialties. Another likely separation was into supposedly practical work and that far removed from use; pure and applied science were born.

Various specialists in science have in private promoted their own or another branch of science as basic and important, ranking others on a lower level. Sometimes the classification is friendly jesting, but it has erupted into serious confrontations between the physical chemist and organic chemist or the physiologist and taxonomist. The situation between those in applied science and what is called pure science was an imperceptible ripple until the last half of the twentieth century. Funds were no longer available to support both in style, and the excesses of technology were making enemies rather than friends. Differences were sought and aired.

In 1879, Henry A. Rowland at Johns Hopkins University said: "He who makes two blades of grass grow where one grew before might be a benefactor of society, but he who labors in obscurity to find the laws of such growth is the intellectual superior as well as the greater benefactor of the two." Such a concept has more or less been promoted by the advocates of so-called pure science. For example, chemist and former presidential science advisor George B. Kistiakowsky wrote:

> The point that requires repeated emphasis is that closely-defined mission-oriented research has value but, by itself, is insufficient and incapable of developing really new ideas and new principles on which each practical mission will ultimately find itself based. If the social climate and support mechanisms are not such as to encourage the free exploration of new ideas rapidly and effectively our technology will die on the vine because in the absence of the results of new, undirected basic research applied work tends to become more and more confined to increasingly expensive refinements of and elaboration of old ideas.

Not all disinterested observers agree with that contention. In 1965, the National Academy of Sciences issued a report, *Basic Research and National Goals* containing a chapter titled "Federal Support of Basic Research" by Harry G. Johnson, professor of economics at the University of Chicago. He wrote: "In order to establish a case for Government support, it must be shown that basic research yields a social return over its cost that exceeds the return on alternative types of investment of resources." He concluded that information was lacking to establish the case. Others closer to natural science have expressed similar sentiments. In 1966, Ivan Bennett, the Deputy Director of the Office of Science and Technology, said: "Science . . . can no longer hope to exist, among all human enterprises, through some mystique, without constraints or scrutiny in terms of national goals, and isolated from the competition for allocation of resources which are finite."

Just as pure science has its promoters, applied science has its boosters. The point of view of Henrik Bode, an engineer at Bell Telephone Laboratories, is indicative. He claims that

> Science was far more indebted to technology than technology to science . . . the invention of both the telescope and the microscope depended on a flourishing industry in spectacle lenses that already existed. Magnetism was known as an empirical fact, and had been used as a basis for the navigator's compass for centuries before the eighteenth and nineteenth century physicists got around to studying the phenomenon.

Watt's steam engine was invented without the benefit of the Carnot cycle, or Joule's work.

In the same vein, J. Herbert Holloman, an engineer appointed by President Kennedy to be Assistant Secretary of Commerce for Science and Technology, said in 1966 at a lecture at Washington University: "New technology flows from old technology, not from science."

The controversy about the relative values and interrelationships of pure and applied science has been almost entirely favorable to technology. In 1963, the Department of Defense selected twenty crucial weapons and analyzed whether an original scientific principle propounded since 1945 was the basis for each item. The conclusion in *First Report on Project Hindsight* (June 30, 1966) was: "In the systems which we studied, the contributions from recent undirected research in science was very small." They did admit that "without the organized body of physical science extant in 1930—classical mechanics, quantum mechanics, relativity, thermodynamics, optics, electromagnetic theory, and mathematics—none of our applied science and only a fraction of technology events could have occurred." Pure scientists reacted as expected. They were somewhat gratified by a report issued in 1969 by the Illinois Institute of Technology with National Science Foundation sponsorship, *Technology in Retrospect and Critical Events in Science.* The document traced five important technical innovations to research done long ago with little or no thought of future utility.

In 1973, the National Science Foundation released the report "Interactions of Science and Technology in the Innovative Process: Some Case Studies" done by Batelle Columbus Laboratories. The investigators studied oral contraceptives, magnetic ferrites, and the video tape recorder (also a subject of the 1969 Illinois Tech paper) the heart pacemaker, hybrid grains, electrophotography, input-output economic analysis, and organophosphorus insecticides. The achievements were traced to a reservoir of basic knowledge, some attained as long ago as 1790. They also highlighted the "technical entrepreneur" as a dominating force for innovation, pursuing his objective in the face of difficulties and discouragement. Also in 1973, the Department of Liberal Studies of Manchester University, England, issued the result of a two-year study demonstrating that "information which results from scientific activity is important to innovation." The investigators chose a set of thirty technical advances and traced back the origins of the ideas.

The trend was, however, in favor of engineering and technology. In 1967, the National Academy of Sciences issued a collection of sixteen essays entitled *Applied Science and Technological Progress.* In 1964, the National Academy of Engineering was established under the charter of the

National Academy of Scientists, and the strain that developed between the two groups was more than the problems of early adjustment. By 1970, the National Academy of Sciences had eight hundred and seventy members and the National Academy of Engineering had grown to three hundred and twenty-seven members.

Lyndon Johnson, as president of the United States, described the drive toward applied science with: "A great deal of basic research has been done. I have been participating in the appropriations for years in the field. But I think now the time has come to zero in on the targets by trying to get our knowledge fully applied." In 1971, economist James R. Schlesinger, appointed to head the Atomic Energy Commission, expressed the sentiment another way when he said: "There has been an atmosphere in the labs wherein a researcher who doesn't want to work on an assigned task doesn't have to—that this is all part of the spirit of free inquiry. Well, we can't automatically permit researchers to follow this proclivity."

During 1956-64, federal support, mainly military, for basic research grew about 29 percent a year; during 1964-69, its growth rate was only 9 percent a year. The National Science Foundation Amendment Act of 1968 (Public Law 90-407) dictated greater involvement of the National Science Foundation with applied research. Later, the Mansfield amendment to a Department of Defense appropriation bill prohibited money for research unrelated to the military mission. In 1970, for the first time since the establishment of the office, the presidential science advisor was a technologist.

The National Science Foundation, a bastion for basic research, slowly edged into sponsoring applied research but now spends more than 10 percent of its budget for the purpose. In 1969, it began with a program called Interdisciplinary Research Relevant to the Problems of Society. Soon the name was changed to Research Applied to National Needs, and the program was fractioned into four divisions: Advanced Technology Applications, Environmental Systems and Resources, Social Systems and Human Resources, and Exploratory Research and Problem Assessment.

The situation is the same in all advanced countries. At Britain's Atomic Energy Research Establishment, Harwell, general industrial research is being done on a contract basis. That country's Science Research Council report for 1967-68 stated: "The Council knows that more scientists could profitably be engaged in fundamental research but the number available is limited and it is essential to the country's prosperity that a greater proportion, especially of the most able, should be applying their skills to tackling the country's industrial problems."

At the general meeting of the U.S.S.R. Academy of Science in 1968, a proposal was made that the Academy do certain projects in applied science. In Israel, pure scientists are helping to develop the country's industry, and the three oldest academic institutions have established industrial centers. Even analysts of ancient civilizations cited the importance of applied science. According to J. Eric S. Thompson, the decline of the Mayas was due to the preoccupation of the priests with such subjects as mathematics, astronomy, and the calculation of time, in which they led the world, and to the failure to attend to the more practical needs of the farmers.

The apparent victory of applied science is an illusion. Only specious and so-called common sense definitions separate it from pure science. The criterion of utility, widely accepted, is subject to interpretation and misinterpretation. Expediency, economy, and immediacy are in the same flexible category. Thus the king of Denmark spent a fortune supporting Tycho Brahe in an island observatory, and the work was purportedly basic research. Yet the prestige accruing to the king and Denmark and the data retrieved for later use by Johannes Kepler were immensely practical. Newton's laws of motion may have been considered esoteric by some thinkers, when it was first announced, but within a couple of generations, engineers held the concept in great reverence. Pure scientists have continually repeated these arguments, but there are more important reasons why the separation of the two areas has become a myth.

In 1921, English physicist Norman Campbell wrote:

> Pure and applied science are the roots and branches of the tree of experimental knowledge; theory and practice are inseparably interwoven, and cannot be torn asunder without grave injury to both. The intellectual and the material health of society depend on the maintenance of their close connection. A few years ago there was a tendency for true science to be confined to the laboratory, for its students to become thin-blooded, deprived of the invigorating air of industrial life, while industry wilted from neglect. Today there are perhaps some signs of an extravagant reaction; industrial science receives all the support and all the attention, while the universities, the nursing mothers of all science and all learning, are left to starve. The danger of rushing from one extreme to another will not be avoided until there is a general consciousness of what science means, both as a source of intellectual satisfaction and as a means of attaining material desires.

Michael Ference, who spent ten years in the physics department of the University of Chicago before working at the Ford Motor Company's scientific laboratory for more than two decades, believes science is a unity covering pure and applied sectors. Harvey Brooks, professor of applied physics at Harvard University and a leader in American science councils, believes "as definite categories, basic and applied tend to be meaningless, but as positions on a scale within a given environment they probably do have some significance." Sir Solly Zuckerman wrote in his 1961 report *The Management and Control of Research and Development:* "There is and can be no clear-cut line of demarcation between one form of research and another; basic research and development are, so to speak, bands at opposite ends of a continuous spectrum." Israeli sociologist Joseph Ben-David in his 1968 report published by the Organization for Economic Cooperation and Development (OECD) wrote: "There is no reason why some applied research should not be also fundamental research and *vice versa.* If a piece of research—whatever its original aim—results in publication which is a contribution to knowledge and also leads to the solution of a practical problem, then it is both." In *Scientists in Organizations,* (New York: John Wiley, 1968), Donald C. Pelz and Frank M. Andrews, studying thirteen hundred scientists and engineers, report that effective scientists do not limit their activities either to the applications world or to that of pure science; they maintain an active interest in both, and their work is diversified. Political scientist Michael D. Reagan of the University of California, Riverside, a student of the scientific community, wrote: "What is basic can be differentiated not by the motives of the investigator, nor by the applicability or significance of the result in any direct sense, but by the *conditions* under which the research is done." Conditions of work vary from organization to organization so that what might be basic research by an outsider's definition could be applied research to the affected individual. In 1965, some astronomers at the National Radio Astronomy Observatory, Green Bank, West Virginia, were dissatisfied with the living accommodations, so they moved to a branch office on the campus of the University of Virginia in Charlottesville, one hundred miles away. Seemingly a pure science establishment, the observatory had for them the characteristics of an applied science organization.

The similarites between the two fields are much more numerous than the differences. Both use the procedure of isolating a problem within narrow limits and verifying hypotheses for solution through experiment and empiricism. Case studies, field investigation, and statistical analyses are probably used in the two areas with the same degree of frequency. Through habit, the so-called pure scientist may invoke the loftier word *research,* and those designated as applied scientist may be more familiar with the title *development,* although their modes of operation do not vary. Empiricism—minimum recourse to theory and seeking a practical result—is no doubt employed as much by those in the university laboratory as it is by those in the industrial research center. Lord Rothschild, a biologist and head of England's Central Policy Review Staff, reported in 1972 that basic and applied research are indistinguishable as to how they are done. According to him, "the distinction lies in why research is done and who wants it done." Commenting on the Rothschild report, critic Eric Ashby wrote:

> In pure research his goal is simply to solve some problem *within* the discipline (chemistry, biology or whatever). If he succeeds, he will get a solution valid only within the discipline and this will satisfy him; it is all he wants. In applied research he has a double purpose: to solve some problem *outside* the discipline (usually with a social goal: to cure disease, to increase productivity, to improve transport) and in doing so, to use the methods and techniques acceptable within the discipline.

The practical man may believe that applied research more often leads to successful conclusions but the facts to support the contention are not available. The degree of failure for scientific projects may be as high as it is for technological ones. The latter, like the work of psychoanalysts, is not usually open to inspection. Corporations as well as pure research institutes do not publicize aborted efforts and inquiries not coming to fruition.

Laboratories of any kind are populated by those more interested in ideas and things than people; greater variations in human characteristics occur inside and outside the subjects than between those who do pure and applied scientific work. The usual consensus that engineers are more conservative politically may be only a result of guilt by association. Technologists are not self-employed or in citadels of thought that universities are supposed to be; they are with the establishment and may be politically oriented by osmosis.

Pure and applied scientists are funded by public money. The bulk of research expenditure in every country is by the government, even though corporations in the United States advertise their zeal for supporting scientific work. The money comes from the same spigot, but the control of it—that is, the conditions under which used—varies.

If the separation between pure and applied science is a myth, the disputes about who gets more funds and who is more significant should, in time, become less virulent. In 1960, astronomer Fred Hoyle wrote: "Technology is a little like strong liquor. A little won't do you much harm, but too much is only too apt to drive every sensible idea out of your head. So far as technology is concerned it is my opinion that our modern society is dead drunk." Such strong condemnation also emanates from the new conservationists aroused by the pollution and pillage of our planet. The ecology movement will be a strong factor in bringing pure and applied science back to the same perspective. One of the "purists'" credos—that everything is related—should return unity to the house of science. The point was best expressed by Russian A. S. Emelyanov at the tenth

Pugwash Conference. He said: "A scientist cannot be a 'pure' mathematician, biophysicist or sociologist for he cannot remain indifferent to the fruits of his work, to whether they will be useful or harmful to mankind. An indifferent attitude as to whether people will be better or worse off as a result of scientific achievement is cynicism, if not a crime."

The ecology concern is sooner or later an international one, and this may lead to financing pure and applied science on a planetary rather than a national basis. An international body in charge, as was outlined at the end of Chapter 1, is in line with the international character of the discipline despite the fact that the pioneers and developers of science were largely Western European. Social commentators such as historian Arnold Toynbee have noted a greater international brotherhood among scientists than among any other group, including political proselyters.

SELECTED READINGS

Goran, M. "The Marriage and Divorce of Pure and Applied Science," *The Chemist*, 48 (April 1971), 107-108.

Klaw, Spencer. *The New Brahmins, Scientific Life In America* (New York: Morrow, 1968).

National Academy of Sciences. *Applied Science and Technological Progress* (Washington, D. C.: Government Printing Office, 1967).

National Academy of Sciences. *Basic Research and National Goals* (Washington, D.C.: Government Printing Office, 1965).

Reagan, Michael D. "Basic and Applied Research: A Meaningful Distinction?" *Science*, 155 (March 7, 1967), 1383-1386: 159 (January 19, 1968), 255.

Technology and Culture, v-6, #4, 1965.

Vollmer, Howard M. "Basic and Applied Research," in Saad Z. Nagi and Ronald G. Corwin, (Eds.), *The Social Contexts of Research* (New York: Wiley, 1972).

Weinberg, Alvin M. *Reflections On Big Science* (Cambridge, Mass.: MIT Press, 1967).

CHAPTER 10

Certainty in Science

The view of science from a distance, comparable to the pedestrian look at religion, business, or government, yields a blur encompassing many diversities. No too long ago, advertisers eagerly embraced the word "science" to give their products an aura of certainty and progress. Half of this tragedy was the improper presentation and exploitation of science, but the other part, perhaps more serious, is that much of the scientific community equates science with certainty. The myth is perpetuated in several ways.

For one, every science has an array of material from established fact to educated guess, yet all may be given to the student with equal time and emphasis. In astronomy, the size, shape, and motions of the earth are known with more certainty than the distances of galaxies, but both may be taught without attached provisions. In geology, landscape features are a matter of definition and citation, while the conception of plate tectonics and sea-floor spreading is a relatively new idea. Yet the two, varying in significance, may have a week of classtime apiece. Only rarely does an instructor show the spectrum of certainty in a subject.

All sciences have a tendency to package data and concepts neatly, without loose ends. The results are boxes containing the uncertain as well as certain. If biologists need the categories of plant and animal, but no other, wherever *Euglena* is placed has a questionable element. If life is carbon based, then the classification of radiolaria with its silicon introduces an item of uncertainty.

Every science has areas where pioneering is done. Exciting to the research participants, those areas are useful as devices for the accumulation of funds and are glamourized and better advertised than the more established parts of the subject. The nonspecialist may come to believe that genetic engineering for human beings or space travel to the stars is as certain as currently used applications of science.

All sciences employ the doctrine of the uniformity of nature with its inherent uncertainty. The contention that identical or similar situations bring like or almost like results is a prime belief among all thinking men and women, nonscientists as well as scientists. Yet, the idea of uniformity can be misleading. The surface of the moon not seen until 1959 was presumed to be much the same as the surface facing the earth; photographs taken by the astronauts show differences. Again, any sample of a radioactive isotope contains billions of atoms identical in composition and in structure and their behavior should be identical, but some decay spontaneously and others do not. Organic evolution proceeds because a few organisms in every species turn out differently from their parents, even though the conditions of their embryonic development appear to be like those of their undifferentiated siblings. Fortunately, scientists recognize that nature is diverse as well as uniform, and they pay strict attention to exceptions and apparent negative cases. The so-called certain information yielded through application of the concept of uniformity of nature cannot be ascertained unless it is thoroughly corroborated.

One application of the uniformity of nature principle that too often leads to difficulties is the tendency toward reductionism. An example is the indiscriminate application of the doctrine of evolution to people, societies, or industries when, in fact, it has been shown to be true only for organisms. Another case leading to error is the exclusion of all concepts from biology save the principles of physics and chemistry. The tendency of some scientists to reduce biology to physics, the person to a series of behaviors, or minds to machines does not make for certainty.

Experiments, too, can be misleading. They test only the consequences of a generalization, not the generalization itself. If an investigator believes that life exists on Mars, he can try to corroborate the idea in many ways. One is to search for evidence of the plant pigment, chlorophyll, since this chemical is found in green plant life on earth. Should the search prove to be fruitless, the original generalization does not fall asunder. One swallow does not make a summer, and one or two positive—or, for that matter, negative—results do not yield a fully substantiated idea.

Not all subjects classified as science have equal certainty. Medicine, for example, is considered as a science by many. But John C. Whitehorn, professor emeritus of psychiatry at John Hopkins University, in his lecture, "Education for Uncertainty," at Massachusetts General Hospital in 1961, said: "The dogmatic assumption of determinism in human behavior, fostered in large part by the sophomoric expectation of certainty in knowledge has a pernicious effect on the practice of medicine, by leading the young doctor to try compulsively by the unwise and neurotic multiplication of tests and superfluous instrumentation to achieve the illusion of certainty—a modernistic and expensive superstition."

Psychiatry, too, has often been criticized for its lack of certainty. The letters section of *Science* (February 10, 1967) has such a discussion by seven writers. Psychology has been labeled in the same way. Abraham H. Maslow, in his *The Psychology of Science* (Harper and Row, 1966), wrote:

> Most young psychologists have been taught to use the controlled
> experiment as the model way of acquiring knowledge. Slowly

and painfully we psychologists have had to learn to become good clinical or naturalistic observers, to wait and watch and listen patiently, to keep our hands off, to refrain from being too active and brusque, too interfering and controlling, and—most important of all in trying to understand another person—to keep our mouths shut and our eyes and ears wide open.

This is different from the model ways in which we approach physical objects, i.e., manipulating them, poking at them, to see what happens, taking them apart, etc. If you do this to human beings, you *won't* get to know them. They won't *want* you to know them. They won't *let* you know them. Our interfering makes knowledge less likely, at least at the beginning. Only when we already know a great deal can we become more active, more probing, more demanding—in a word, more experimental.

Maslow's brand of psychology disagrees with many interpretations of the subject, just as different schools of thought exist in other sciences. Certainty of knowledge become illusory when humanistic, gestalt, depth, and other views of the same facts are available. Indeed, facts may turn into theory and theory into facts when new approaches are used.

Changes are most noticeable in a single generation when a science is young, but in the history of every science are records of revolutionary transformations. How certain could a consumer of scientific knowledge be during the several hundred years when the Ptolemaic geocentric view of the heavens was being supplanted by the Copernican heliocentric concept? What certainty could thinkers outside of chemistry have when the phlogiston idea was cast aside? Physicists took sides about caloric and ether, but could a nonphysicist be certain when these interpretations were being contested?

Facts dispersed by science do not always dispel uncertainty. For example, during the twentieth century, the age of the earth has been described as being two billion, three billion, and four and one-third billion years. The distance to the Andromeda galaxy changed from eight-hundred thousand to two million light years. The new values can be ascribed to better refinements and closer approximations to the truth, but some factual changes can affect human lives in a deleterious manner. Early in the twentieth century, the mobile chest x-ray van was promoted as a great aid for finding active cases of tuberculosis. By 1971, the National Tuberculosis and Respiratory Disease Association urged their abandonment because the devices exposed subjects least likely to have tuberculosis to needless radiation.

The medical sciences are particularly prone to generate confusion rather than certainty. In one generation, child-rearing theories may fluctuate from emphasizing permissiveness to rigid discipline and back to permissiveness; pregnant women may be advised to become fat or remain as thin as possible; panaceas such as vitamin E, prostaglandins, and vitamin C may be alternately promoted and condemned.

Scientists depending on others for information are often laden with uncertainty about the facts. The September 6, 1965, issue of *Chemical and Engineering News* carries a plea from a reader:

> Who has not had the frustrating experience of attempting to repeat an experimental procedure from the technical literature without encountering problems not spelled out in the experimental section of the paper? . . . If a scientific publication is to perform the useful function of helping other workers in the field, and not merely serve the purpose of bringing fame and glory to its author, then it is essential that all the useful hints and tricks be imparted to the interested reader.

According to the view of English scientist P. B. Medawar, the scientific paper in general is fraudulent because it misrepresents scientific thought. He would discard the inductive format and have scientists describe the hypotheses that occur to them along the way of investigation.

Dr. Neal E. Miller, professor of physiological psychology at Rockefeller University and a former president of the American Psychological Association, told a congress in Moscow:

> Published reports of research are written with the wisdom of hindsight. They leave out the initial blind groping and fumbling to save journal space (and perhaps also to save face), and exclude almost all of those attempts that are abandoned as failures. Therefore, they present a misleading picture, which is far too orderly and simple, of the actual process of trying to extend the frontiers of science into unknown territory.

What and when to publish is a problem for many research workers. They must present themselves in the best possible way to their peers and often what is published will depend on the scientist's age, position, ego, and sense of security. A young person may simply want to establish his presence and possible priority without describing something completely that has not yet been thoroughly checked; a middle-aged worker may rush into print in an effort to maintain status; the most mature scientist may omit those facts that seem self-evident to him, but they may not be so to many of his readers.

The scientist seeking certainty in the literature of his subject must wade through mountains of material and use his judgment in sorting useful from useless. The proliferation of scientific publication is castigated by most public and private commentators on the topic, but the problem remains with no solution on the horizon. During the summer of 1964, many letter writers to *Chemical and Engineering News* complained about the corruption of scientific literature with publications of little or no value. A man with twenty years of experience in government laboratories complained in the December 1966 issue of *Scientific Research* that the "publish or perish" dictum promotes the deluge of material.

One of the most decisive criticisms of certainty in science comes from widely published material on the subject of nuclear physics. Its principle of uncertainty proclaims the inability of man to measure precisely both the position and speed of subatomic particles. In this realm, the doctrine of causality begins to fall asunder.

The presence in science of educated guess as well as substantiated fact, of less-established subjects as well as older accepted materials, of changing ideas and facts as well as the time-tested, of important published data as well as space fillers, and Heisenberg's principle of uncertainty demeans the claim of certainty.

However, when compared to other interpreters of nature, past and present, certainty according to science begins to shine.

In the Middle Ages, the Doctrine of Signatures purported natural phenomena held messages for man; accordingly, herbs were thought to be stamped by nature with a clear indication of their curative properties. The blood-red beet was good for the heart, the upper parts of plants were for treatment of the upper regions of the body, and perennial plants gave long life. Mistakes have occurred in today's pharmacology, but the overall thrust has been more beneficial to man's health. In a similar manner, astrology purports to forecast human affairs on the supposition of a relationship between date of birth and star patterns. Aside from the entertainment provided astrology has been of limited value—certainly of much less value than the meager predictions that can be made from the modern science view, pointing to heredity and environment as indicators.

SELECTED READINGS

Bridgman, Percy W.   "The New Vision of Science," *Harper's Magazine,* **158** (1929), 443-451.

Dewey, John.   *The Quest for Certainty; A Study of the Relation of Knowledge and Action* (New York: Minton Balch, 1929).

Grene, Marjorie (Ed.).   *Interpretations of Life and Mind: Essays around the Problem of Reduction* (New York: Humanities, 1971).

Medawar, P. B.   "Is the Scientific Paper Fraudulent?" *Scientific Research* (August 1, 1964), 19-20.

PART THREE

Surprises

Duration time is one difference between myth and surprise. Myth can last through several generations; surprise is an event bringing more immediate awareness. Consequently, one may shade into the other. For example, today, many literate men and women are surprised to find that, long before Christopher Columbus, thoughtful people believed the earth to be spherical. During the second century BC, Eratosthenes made the first successful measurement of the circumference of the earth, and he necessarily assumed the spherical shape. The idea that Columbus was the first to indicate the sphericity of the earth is a myth, but when confronted with the evidence, the reaction of people is surprise.

Tragedy, comedy, and discovery are in this category of evoking surprise, but not to the alerted observers. When death and disease strike, they are tragedies to those affected although the analyst armed with statistics can claim he knew what would happen. The pun, the mistaken identity, or any comedy provoking laughter can also be ascertained beforehand, yet to those who do not envision the unexpected, they are surprises. Discovery and invention seem to be in this category; although popular expositions of the events would have them be a surprise of the first magnitude, closer examination reveals enough clues and necessary prior activity to discern a trend toward discovery. When the nature of atomic structure was first probed at the beginning of the century, the results were surprising to the uninitiated but not to those actively engaged in research in the area. Very few voices of dissent were raised to the atomic interpretations of J. J. Thomson and Ernest Rutherford.

The great names of science are usually associated with surprising discoveries or, in modern parlance, breakthroughs. They are surprises to the novice but not to the expert. Thus, in a preliminary survey of the history of biology, a huge vacancy occurs between the accomplishments of Galen in the second century and those of Vesalius in the sixteenth; closer study reveals a number of advances in the generations before Vesalius.

Science and scientists do have their surprises, even to those in the field. Some of these are presented in the next two chapters.

CHAPTER 11

Surprises about Scientists

Late twentieth-century science bears very little resemblance to the science of Galileo's, Newton's, and Darwin's times. Even in 1940, the tenor of the laboratory was significantly different than it is today; then, the scientific community was small. In the United States, the suggestion of federal support for research had overtones of heresy. Yet when many nonscientists think about science and scientists they conjure up images of the pioneers; under the circumstances, surprises are frequent.

Until the profession of scientist developed at the beginning of the twentieth century, the pioneers of science were largely amateurs. Surprisingly, many amateurs are still making contributions beyond joining rock-collecting or star-gazing clubs, and their achievements are outstanding in some cases. The godfather of artificial rainmaking, Vincent J. Schaefer, was trained as a forester, without an academic degree, and worked on wartime gas masks in the General Electric Laboratories in Schenectady in 1946. The distinguished thinker and inventor R. Buckminster Fuller, was twice dismissed as a student by Harvard University. More often the amateurs have a minor albeit supporting role such as that of farmer Masaji Nagata of Brawley, California. He found a comet now known by his name.

Amateurs today do not usually have the support of the professional science establishment. Lt. William Fox of the New York Police Department, a PhD chemist, failed to acquire government money for his research when others of similar background but professionally affiliated had little difficulty in obtaining federal grants. Frank Restle of the Department of Psychology, Indiana University, describing the work of an amateur astronomer in the June 12, 1970 issue of *Science,* wrote: "It is frightening to consider that an idea may be unpublishable when offered by a comparative unknown, but published in a very prominent journal when offered by a professor of the correct department from a respectable university. We must be alert to open science to the insightful amateur— especially when we consider what a large fraction of the literature published by professionals is derivative and how little can really be said to break new intellectual ground."

Twentieth-century science is characterized by large installations for research, with team efforts in discovery and research. The pioneers worked alone, but, surprisingly, the lone inventor is still with us. A recent one is Stanford R. Ovishinsky, a self-educated high-school graduate who formed his own electronics company, Energy Conversion Devices, Inc. Another who has his own company is physical chemist Norman Haber, who acquired all his higher education at night school.

Now science has institutions and curricula for the training of scientists; the pioneers were largely self-educated. The self-educated are still numerous; among those who have graced our era are the inventors Thomas Edison, Elmer Sperry, Edward Acheson, and George Eastman Physicist Oliver Heaviside never went past elementary school, and astronomer Edward Barnard had only two months of formal schooling. Today's scientists may have university degrees in one subject and learn another themselves; thusly, biochemist Joseph Needham became the western expert in early Chinese science. Einar Hertzsprung was a chemical engineer before becoming an astronomer, and astronomer Carl Sagan was first a biologist. Physical chemist Fritz Haber claimed to be self-educated despite his extensive education in chemistry. Charles Darwin went to a medical as well as theology school; he said, "I consider all that I have learnt of any value has been self-taught."

Many pioneers came to scientific research after training in medicine, law, or theology. The first recognized student of comparative animal behavior, Lewis Henry Morgan, was a lawyer. Among the pioneers who had medical training were Nicolas Copernicus, Agricola, a founder of metallurgy, mathematician Jerome Cardin, naturalist Konrad Gesner, botanist Leonhard Fuchs, physicist William Gilbert, chemist Georg Stahl, geologist James Hutton, physicist Thomas Young, chemist Jons Berzelius, anthropologist Johann Blumenbach, and astronomer Heinrich Olbers. The trend is still with us. Astronomer Edwin Hubble was first a Rhodes scholar in law at Oxford University. Physicist Prince Louis Victor de Broglie was a historian at the start of his career. Anthropologists Bronislaw Malinowski and Franz Boas earned doctorates in physics. Child psychologist Arnold Gesell and psychoanalyst Sigmund Freud were trained as physicians. American expert on meteorites Harvey H. Nininger first taught biology in a Kansas college. Electrical engineer Frank A. Pertet turned to the study of volcanoes. At the Salk Institute in San Diego, Edwin Lennox was a nuclear physicist before becoming an immunologist and Jacob Bronowski was a mathematician prior to his career in writing and administration.

The move in the other direction, away from science, also occurs. During the seventeenth century, Blaise Pascal finally turned to religion. Recently, black chemist George A. Wiley founded and headed the National Welfare Rights Organization.

The "natural" scientists—the amateurs, the self-educated, and those who changed their fields of endeavor—must have been diligent workers in order to have their achievements recognized by their peers. They were intensely dedicated, spending practically all their time with scientific research. Such total absorption characterized many early scientists, well known and not. Isaac Newton was once described: "I never knew him to take any recreation or pastime either in riding out to take the air, walking, bowling, or any other exercise whatever, thinking all hours lost that was not spent in his studies to which he kept so close that he seldom left his chamber except at term time." A physicist of lesser stature, Simeon Poisson, has been similarly described: "As a student at the Ecole Polytechnique he had worked particularly hard to the exclusion of social activities. Later, particularly after his marriage in 1817, he received no visitors but worked in his study from early morning until dinner at six." The same kind of hard worker can be found in science today. In 1973, Dean Lewis Thomas of Yale University (now president of the Memorial Sloan-Kettering Cancer Center in New York City) wrote about "natural" scientists: "I don't know of any other human occupation, even what I have seen of art, in which the people engaged in it are so caught up, so totally preoccupied, so driven beyond their strength and resources." Physicist-writer Charles Percy Snow has described a typical case in his *The Search*. The prinicipal character is told by a high school instructor that research would mean years of work, even preventing marriage.

Unlike those in other professions, many of the pioneers of science never married. The list of the great names who were bachelors includes Newton, Pascal, Boyle, Huggins, and Cavendish. The latter was a prize catch for a woman who wanted riches since he was one of the wealthiest men in England of his time. Henry Cavendish, however, avoided social contact.

The root cause of remaining a bachelor could be absorption in work, but a deviant family situation can also be involved. Many pioneers as well as current scientists lost one parent early in life. These include Newton, Kelvin, Lavoisier, Boyle, Huggins, Rumford, Curie, Maxwell, Copernicus, Brahe, Descartes, and Priestley. The loss of a parent can trigger poor interpersonal relationships (such as Henry Cavendish had), failure to marry, and other unusual characteristics.

Newton was one prototype of today's laboratory people who have unique personalities. Economist John Maynard Keynes had an unflattering description of Newton: "For in vulgar modern terms Newton was profoundly neurotic of a not unfamiliar type, but—I should say from the records—a most extreme example. His deepest instincts were occult, esoteric, semantic—with a profound shrinking from the world, a paralyzing fear of exposing his thoughts, his beliefs, his discoveries in all nakedness to the inspection and criticism of the world." On the other hand, many of the other pioneers were extroverts. Brahe, Galileo, Hooke, Lavoisier, Volta, Davy, and Liebig could be leading lights in any company.

Such men and women are in laboratories today. Astronomer Charles A. Whitney, assessing his teacher, astronomer Harlow Shapley, wrote: "I have never seen a quicker mind, a more agile sense of humor, or a more complete absence of what usually passes for humility." Physicist Robert Serber describing his teacher, physicist J. Robert Oppenheimer, said: "The world of good food, good wines and gracious living was far from the experience of many of them, and Oppie was introducing them to an unfamiliar way of life. We acquired something of his tastes. We went to concerts together and listened to chamber music. Oppie and Arn Nordsieck read Plato in the original Greek. During many evening parties we drank, talked and danced until late."

Mathematician John von Neumann was described by a close co-worker at Princeton University: "He . . . greatly enjoyed people's company, and his house was the scene of most wonderful parties and dinners . . . at least once a week he and his wife would entertain a houseful of people."

Another remarkable person of our day is Russell Marker. After receiving his degree in chemistry from the University of Maryland in 1923, he studied for a doctorate for two years but left following an argument with one of his professors. He worked at the Rockefeller Institute for Medical Research and left that institution after a minor disagreement with a fellow worker. He quit his job at Pennsylvania State University in the middle of a school term and went to Mexico. In 1943, he and two others founded an organization for the preparation of sex hormones. In 1946, he sold his interest after a dispute and started a new company. He left, six years later, and spent the remainder of his life in retirement from science. Marker was obviously not a team personality.

Current arrangements for research and development bring many men together as a team. The reports in journals and patent offices sometimes contain a long list of authors. The arrangement spells cooperation as to exchange of information and subdivision of the problem into parts. Yet, surprising as it may seem, the modern way of doing things has not diminished the creative role of the individual. Long ago, solutions were found by individuals, and such is the case today in many instances. Not all who are outstanding in their abilities have positions of group leaders, research directors, or full professors.

The genius in physics of the nineteenth century was Michael Faraday, said to "smell the truth." Twentieth-century scientists, even though they have most often been engaged in group endeavor, are similarly described. The father of nuclear science, Ernest Rutherford was said to know by instinct what observations were important and what were trivial.

Not every laboratory scientist of stature can be equated to an efficient worker; some were very poor experimenters. Chemist Humphry Davy, in the early nineteenth century, was careless and slovenly. At about the same time, chemist Robert Bunsen had fingers that operated like thumbs.

Of course, not all members of the profession are geniuses. One of the first scientist members of the United States Atomic Energy Commission confessed that the most difficult part of his job was to make a decision before acquiring all the evidence, a feat which is done by all top-notch scientists.

The creative role of the individual was disparaged long ago and, surprising as it may seem, is downgraded today, even with our greater knowledge of how the mind operates. Lucky accident or chance was the usual description of the event. Such was the appellation for the vulcanization of rubber by Charles Goodyear who retaliated with, "While I admit that these discoveries of mine were not the result of scientific chemical investigation, I am not willing to admit that they were the result of what is commonly called accident. I claim them to be the result of the closest application and observation." In our time many an achievement has been labeled a lucky accident. French physicist Henri Becquerel put an exposed photographic plate in a desk drawer containing uranium ore. Not everyone would have interpreted the result as did Becquerel; he saw that the photographic plate must have been affected by something coming from the uranium ore, and he was led to the discovery of radioactivity. When Percy W. Bridgman was a twenty-three-year-old graduate at Harvard University in 1905, a high-pressure chamber for an optical experiment broke. While waiting for a replacement part from Europe, he devised a joint for high-pressure that gave no trouble. He saw immediately how leak-proof joints and fittings opened up a new field for high-pressure investigations. Obviously, Louis Pasteur's dictum "chance favors the prepared mind" is still valid.

Perhaps a lucky accident, serendipity is where a product, process, or idea more valuable than one originally sought is obtained. It is a surprise to the investigator and to the entire scientific community. Examples are legion in every subject. In chemistry, a purple aniline dye was found by William Henry Perkin while he was trying to synthesize quinine. In nucleonics, Ernest Rutherford found the structure of the atom during the course of his study of the scattering of the radioactive particle, the alpha ray. In physics, Max Planck came upon the quantum idea while trying to account for the radiation from black bodies. In mineralogy, a prospector searching for gold in 1929 at Great Bear Lake in the Northwest Territories, Canada, recognized extensive uranium deposits overlooked by others. The discovery of nuclear fission resulted from studies in the action of neutrons on uranium, supposedly forming transuranic chemical elements.

Serendipity, a positive reaction, has been widely publicized, but the negative response which is a surprise has been comparatively ignored. It, too, occurs in all sciences. At the start of the twentieth century, the chemical element helium was detected in the sun, and some astronomers expected other elements to be found. Analyses of nebulae showed the possibility of one, and nebulium was christened; in 1927, the supposedly new element was shown to be a variety of nitrogen and oxygen. Somewhat earlier, Michael Faraday was involved in a negative surprise in the same science of spectroscopy. He placed a flame of burning sodium between the poles of a magnet and looked unsuccessfully for some influence of the magnetic field on the spectral lines. Thirty years later the effect was found by Dutch physicist Zeeman.

Technology, too, has its negative surprises, with sometimes devastating effect. Any person sensitive to environmental disruption can cite at least one deleterious impact. For the engineers, however, the negative surprises point up overlooked problems. One occurred during the late afternoon of November 9, 1965, when a small electrical relay in a power station in Ontario, Canada, failed. Within a few minutes, the flow of electric energy stopped throughout parts of Canada and much of the northeast section of the United States. Late in 1973, the Environmental Protection Agency began considering whether they should do away with the catalytic converters U.S. automakers planned to install in most of their 1975 model cars. The catalyst was designed to reduce hydrocarbon and carbon monoxide emissions. However, some tests showed that the platinum-lined catalyst was promoting the conversion of sulfur in gasoline into sulfuric acid mist.

Discovery via fraud has also occurred. One of the best cases is the finding of the bark of cinchona as a cure for malaria. At first, the bark of a Peruvian tree known by its native name *quina-quina* was somehow thought to be suitable. The demand for it rose, and the supply became limited quickly. Dishonest merchants began to substitute the bark of another tree, the cinchona, since it closely resembled quina-quina, and was found to be an effective therapy.

If serendipity is not involved, the research results of investigators should not be surprises to them. On the contrary, they were motivated to do the research because of some surprising phenomenon. In theory, their desire to comprehend an anomaly in accepted patterns and concepts usually sparks their curiosity and pushes the work forward. In practice, the results of investigations in science have invariably been surprises to some, scientist and nonscientist alike. From the very start, some intelligent men and women have refused to acknowledge the results obtained by other intelligent men and women. Reason can be on both sides.

Those thinkers who opposed Galileo were not stupid. The view through a telescope was a surprise to them, just as instrumental sightings in later centuries were surprises to large numbers of scientists. In 1915, astronomer W. S. Adams inferred that the companion to Sirius, the first white dwarf known, would have to have a density about seventy-thousand times greater than that of water. Not too long afterward, a book written by a scientist claimed that the measurements of Adams were wrong since such a density was clearly impossible.

The failure of intelligent men and women to accept what appears reasonable to later generations may be attributed to mind sets and preoccupation with one perspective. If a geologist is inculcated with the theory of isostasy, plate tectonics must fight for his attention; wave mechanics is not only an entire, new language but also a different approach for those physicists who know only classical mechanics.

New views in any science bring a revolution that may shake a generation or two of workers in the field. These revisions are not a frequent occurrence for any one science. More likely the scientific worker deals with his subject in the manner of his school days and makes his contribution by adding a measurement here or an observation there, but he usually lives and dies with the general approach in which he was educated.

Professors are a conservative lot in general, and science professors are not an exception. They resist the fundamentally new. In 1879, Marquis de Sautuola and his four-year-old daughter walked through the cave of Altamira in the Cantabrian Mountains of Northern Spain and found the first prehistoric cave paintings. The professors who came to examine the pictures accused the discoverer of being a faker; some of them said that he probably had hired an artist in Madrid in order to be acclaimed as an archaeologist.

The scientists who are associated with revolutionary changes are those who have the audacity to break out of the habit of old views. Isaac Newton said that "no great discovery is ever made without a bold guess." Industrial achievements are not exempt. When receiving the Perkin medal, Dr. Vladimir Haensel of the Universal Oil Products Company said that a scientist in industry must be willing to gamble on his judgment and scientific intuition. The man in charge of research at the Xerox company since 1938 reported: "The members of our team were all gambling on the project. I even mortgaged my house. All I had left was my life insurance. My neck was way out. My feeling was that if it didn't work Wilson and I would be business failures but as far as I was concerned I'd also be a technical failure."

If science is considered to have both revision and accretion aspects, then the surprises come with the former when concepts and ideas are modified or discarded. When the result is not challenging but refines a measurement or adds new data to support an old idea, the element of surprise is much less likely to be present.

The general public associates martyrs of science with the revolutionary ideas, but the true martyrs, surprising as it may seem, are not in the scenario of change. Galileo, for example, did not really suffer compared to those who died in the course of their work. Chemist Berthollet's son died after imbibing too much of the gas he was investigating. Physical chemist Louis Slotin was a hero and a martyr who died when he pried apart fissionable material at Los Alamos, New Mexico, preventing a nuclear holocaust. Edward Jenner did not suffer when introducing vaccination nor did Louis Pasteur when testing an anthrax vaccine. There is no record of Gregor Mendel, the founder of genetics, being adversely affected because he was veritably ignored by the botanists of his day, but there are indications that chemist Scheele shortened his lifespan by testing chemicals.

The general public associates the revolutionary ideas with socially responsible scientists, but again, surprising as it may seem, scientists on the whole, whether involved with revision or accretion, have been socially responsible.

The record from that start shows involvement of scientists with human values. Among the early ones who served governments were Aristotle, Galen, da Vinci, Hevelius, Newton, Boscovich, Lavoisier, and Berthelot. In the nineteenth century, Michael Faraday would have nothing to do with poison gas. During the early twentieth century, Albert Einstein said: "Concern for man himself and his fate must always form the chief interest of all technical endeavors. Never forget this in the midst of your diagrams and equations." At the end of World War I, a group of English scientists banded together for the good of themselves and others, and one of their achievements was the Parliamentary Science Committee to advise their government.

In 1924, American chemist Arthur D. Little wrote: "Science has so drawn the world together and so rapidly remolded civilization that the social structure is now strained at many points . . . Though our civilization is based on science, the scientific method has little place in the making of our laws . . . We see in the ranks of science knowledge without power and in politics power without knowledge."

In the 1930s, Sir Richard Gregory, editor of *Nature,* wrote:

> Men of science, are, however, citizens as well as scientific workers; and they are beginning to realize their special responsibilities for securing that the fruits of scientific knowledge are used for human welfare.They can no longer remain indifferent to the social consequences of discovery and invention, or be silent while they are blamed for increasing powers of production of food supplies, providing means of

superseding manual labor by machines, and discovering substances which can be used for destructive purposes. It would be
a betrayal of the scientific movement if scientific workers failed
to play an active part in solving the social problems which their
contributions to natural knowledge have created . . . Men of
science can no longer stand aside from the social and political
questions involved in the structure which has been built up from
the materials provided by them, and which their discoveries may
be used to destroy. It is their duty to assist in the establishment
of a rational and harmonious social order out of the welter of
human conflict into which the world has been thrown through
the release of uncontrolled sources of industrial production and
of lethal weapons.

When the Nazis in Germany threatened international harmony, both
the American and the British Association for the Advancement of Science
adopted resolutions against them. Physicist P. W. Bridgman shut his
Harvard University laboratory doors to visitors from the totalitarian
states, and well-known American scientists signed manifestos on the
freedom of science, intellectual freedom, and the defense of democracy. In
1938, American chemist Harold Urey said: "I believe I speak for the vast
majority of all scientific men. Our object is not to make jobs and divi-
dends. These are means to an end, merely incidental. We wish to abolish
drudgery, discomfort and want from the lives of men, and bring them
pleasure, comfort, leisure and beauty. Often we are thwarted, but in the
end we will succeed."

Scientists became more active during the late twentieth century, forming a host of societies urging more reason, honesty, and compassion in
running the affairs of man. The release of the atomic bomb set off groups
sponsoring international civil control of the weapon as well as Leo
Szilard's more comprehensive Council for A Livable World. Scientists and
technologists have been in the forefront of the furor about environmental
degradation. Some scientists formed liberal associations: the Federation
of American Scientists, the Scientists Institute for Public Information, the
Society for Social Responsibility in Science. A few began public-interest
science in the Center for the Study of Science in the Public Interest, giving
their full time to such studies as the effects of food additives. Younger
scientists formed Scientists and Engineers for Social and Political Action
and issued a magazine called *Science for The People.* During the early
1970s, virtually every professional science society meeting in the United
States had formal and informal discussions about science and society
topics. For example, on October 30, 1971, Professor Robert H. March of

the University of Wisconsin submitted a petition to the Council of the American Physical Society signed by 276 APS members and proposing an amendment to the APS constitution. Their amendment of professional responsibility said: "The object of the Society shall be the advancement and diffusion of the knowledge of physics in order to increase man's understanding of nature and to contribute to the enhancement of the quality of life for all people. The Society shall assist its members in the pursuit of these humane goals and it shall shun those activities which are judged to contribute harmfully to the welfare of mankind." Another example is the Council for Science and Society, organized in England in 1973. Their objective was to "try to identify areas of research in science and technology which would have important social consequences for good or ill, but which are not yet fully explored; to study these objectively; to attempt to foresee what their consequences might be; whether they could be controlled and how; and to publish responsible reports designed to stimulate wide public debate."

Perhaps the average person associates science with social irresponsibility because more has not been done by scientific groups to stem the tide of greed, selfishness, indolence, stupidity, ignorance, and irrationality. The public, given the technical marvels of jet plane and birth control pill, landings on the moon and the molecule of heredity, also want the miracle of a better man and a better society from the same source—the laboratory.

SELECTED READINGS

Eidusen, B.T. *Scientists* (New York: Basic Books, 1962).

Gilman, W. *Science: USA* (New York: Viking, 1965).

Goran, M. "The Proxy Parents of Pioneer Physical Scientists," *Isis,* **60** (Autumn 1969), 545-546.

Kuhn, T. S. *The Structure of Scientific Revolutions* (2d ed.; Chicago: University of Chicago Press, 1970).

Margenau, H., et al. *The Scientist* (New York: Time-Life, 1965).

Schaar, Bernard E. "Chance Favors the Prepared Mind," Washington: American Chemical Society, Reprint No. 80 from *Chemistry,* 1972.

CHAPTER 12

Surprises about Science

The American colonists were dependent on Western Europe for scientific information. The colonies nurtured a few scientists such as Governor Winthrop and Cotton Mather, and Joseph Priestley crossed the Atlantic Ocean to settle here; but science was not a favored activity in early America. Twentieth-century American sicence is still nourished by Western Europe. The facts are that, between 1949 and 1961, more than forty-three thousand scientists and engineers were admitted to the United States as immigrants; of the forty Nobel prizes awarded to United States citizens between 1907 and 1961, fifteen were won by immigrant scientists; in 1961, one of every six members of the National Academy of Sciences was foreign born.

It is customary to think of science as international, yet until recently, the English, French, and Germans dominated the scene. Americans, Russians, and Japanese are late comers.

Greek contributions to science are inevitably acknowledged, but the association of early Greece with technology surprises many people—even those who hold the evolutionary view of scientific and technological progress and acknowledge that each culture learns from preceding ones. Perhaps those who do not believe in Greek technology conjure up the tale of the tyrant killing the inventor of labor-saving machinery, with the intention of vindicating Greek ability and assigning the supposed lack of applied science to societal pressures.

Every culture has its technology whether or not present-day analysts wish to use the name. Eskimos, for example, have made abundant use of materials in their environment to build a living community. The windows in their igloos are made from the intestinal membrane of the seal or the pericardium of the caribou. Seal and whale oil are used for heating and lighting. Bone, ivory, frozen hides, stone and parts of animal help to sustain and nourish the individual Eskimo. The early Greeks did not have such a challenging situation; their technology dealt less with necessities and more with luxuries.

About 400 BC, geometer Archytas of Tarentum devised mechanical instruments for the description of various curves. Some reports indicate he built an automaton operated through compressed air. The antikythera machine, recovered from the bottom of the Aegean Sea and presumably a clockwork dating from the first century, indicates a tradition in mechanics. A water clock was once in the Tower of the Winds, an octagonal structure built about 50 BC in Athens. The tunnel of Eupalinus, built about 500 BC on the island of Samos, was dug from opposite sides of a mountain to meet in the middle; a true feat of engineering, it was remarkably straight.

Undoubtedly, Greek technology and science were not always successful, just as current research often ends in failure. This fact surprises some laymen, particularly those reared during the first half of the twentieth century, when science was widely presented as *the* problem solver. So-called scientific solutions were the stock in trade of the advertising and political communities. Adding to the euphoria produced by the blind belief in technology as the panacea for all human ills was the tendency to present the history of science as a record of achievements and ignore the concomitant record of failures.

Nowadays, the possibility of failure is discussed in terms of probability rather than as a remote event. The recorder of the first International Conference on Magnetic Resonance in Biological Systems said it more pleasantly in the October 23, 1964 issue, of *Science:*

> One treads the treacherous ground between the hope of dramatic discovery and the danger of having a vast investment of skill, labor and funds rewarded by rather inconsequential observations. It is encouraging to observe, however, that the pure joy of a new adventure which has dominated the field in its first years, is more and more tempered by careful considerations as to whether the questions one is asking are indeed answerable, or, for that matter, worth asking.

Today's increased tempo of scientific activity should bring more success, but the effect may be canceled by the bureaucratic nature of the enterprise. Again, those with false images about science find the bureaucracy surprising. Until the middle of the century and beyond, mass media portrayed the scientist, alone or in groups, as the hero who shows the true way or finds the outstanding product or process. The scientist was never thwarted by a stupid administration or an array of clerks with small empires of authority.

One of the earliest records of bureaucratic bungling has been described by Morton Grosser in his *The Discovery of Neptune.* During the nineteenth century, the astronomer royal in England as well as observatory directors in Europe were too busy dining, napping or socializing to listen to the young astronomers who had pinpointed the new planet.

Examples of current bureaucracy are legion. One with much money is the Special Virus Cancer Program of the National Cancer Institute. Founded in 1964 with ten million dollars, its budget in 1971 was thirty-six million dollars. It has been accused of insulation from the scientific community and concentration of power. In contrast to the system of awarding research grants through review of proposals by peers, the Special Virus Cancer Program distributes its money through contracts negotiated by administrators. It has been accused of sponsoring worthless projects—such as "injecting monkeys with God knows what"—and foolishness.

The last part of that last criticism is, surprising as it may seem, a characteristic of science. Foolishness may be an attribute of our species, permeating all endeavors, including science. Twentieth-century philosopher of science Alfred North Whitehead said, "Almost all really new ideas have a certain aspect of foolishness when they are first produced."

In and out of science, what starts out as foolish can remain that way. Several times during the 1960s astronomers at the Haute-Provence Observatory in southern France reported highly unusual spectroscopic lines from an otherwise normal star; astronomers in California finally pinpointed the anomaly as the light from matches the French used to light their cigarettes. Those who introduced the rabbit into Australia for pest control purposes thought they were doing a good thing; the rabbit became a menace. Similarly, reindeer obstructing railroad tracks in Lapland became a hazard and one thinker suggested installing a loudspeaker emitting wolf calls on the locomotive; the tracks became overrun with wolves.

In and out of science, originators of foolishness seek no recognition—they would rather their efforts be forgotten. On the other hand, positive contributors, in and out of science, want recognition; and it is science that is the most scrupulous of all disciplines in assigning due credit. The author of a poem or the painter of an oil do not inscribe the names of those who taught them; likewise, the novelist and the sculptor make no mention of the sources of their ideas. In pure science and technology, however, mention of prior art is a part of the tradition.

The practice of citing predecessors was not always a feature of science. Copernicus did mention reading about the heliocentric conception in the works of Aristarchus of Samos, but his written reference was scratched out and did not appear in the printed version of his book. René Descartes was indebted to Isaac Beekman and denied the matter. However, by the time the practice of science became a profession, the tradition of mentioning prior art was established.

The man who, in 1930, first devised the cyclotron, Ernest Orlando Lawrence described the origin of his ideas as follows:

> One evening early in 1929 as I was glancing over current periodicals in the University library, I came across an article in a German electrical engineering journal by Wideroe on the

multiple acceleration of positive ions. Not being able to read German easily, I merely looked at the diagrams and photographs of Wideroe's apparatus and from the various figures in the article was able to determine his general approach to the problem—i.e., the multiple acceleration of the positive ions by appropriate application of radio frequency oscillating voltages to a series of cylindrical electrodes in line. This new idea immediately impressed me as the real answer which I had been looking for to the technical problem of accelerating positive ions, and without looking at the article further I then and there made estimates of the general features of a linear accelerator for protons in the energy range above one million volt electrons.

The habit of pointing out previous accomplishers is now so ingrained that any oversight is acknowledged at once. Letters such as the one from Frank Restle, Department of Psychology, Indiana University, in the June 12, 1970, issue of *Science* are not uncommon. Restle wrote:

Among the interesting correspondence which I have received since the publication of my article "Moon Illusion" explained on the basis of relative size (20, Feb., p. 1092) are three articles and a letter from Patrick Rizzo, secretary of the Amateur Astonomers Association of New York. In the articles, published in 1963 in *The Eyepiece,* a monthly bulletin, and a 1960 issue of *Asterisks,* a publication on instructional topics in astronomy, he arrives essentially the conclusion I drew in my article. . . . I acknowledge, his priority in formulating and publishing this relativistic theory of the moon illusion.

Early in 1972, George H. Daniels, a historian of science at Northwestern University, used the same medium to pay his respects to books and authors he had used without giving due credit, to prepare his own work.

The only person to win two Nobel prizes in any science, John Bardeen used his share of the physics prize won in 1972 to set up an endowment fund at Duke University to support the Fritz London Memorial Lectures and Awards. Bardeen said about London: "We are all very grateful to him for the deep insight that helped light the way to understanding."

The tradition of citing prior accomplishment is part of the ethics of science. Somehow the view has arisen that science is amoral and ethically neutral. It is a surprise to nonscientists as well as to many scientists that the institution does have a code of honor. In 1972, the Council of Europe was asked by twenty university professors from fifteen European countries to formulate a code of ethics for science.

The Educational Policies Commission of the National Education Association in their *Education and the Spirit of Science,* issued in 1966, gave one perspective of the ethics in science. Others have cited honesty, cooperation, diligence, open-mindedness, tolerance, international cooperation, and a host of other virtues. In 1953, the American Psychological Association adopted a code of ethics, and in 1973 resolved to follow a set of ten principles governing psychological research with human beings. Biologist Bentley Glass lists four commandments making up the ethic of science: "to cherish complete truthfulness; to avoid self-aggrandizement at the expense of one's fellow-scientist; fearlessly to defend the freedom of scientific inquiry and opinion; and fully to communicate one's findings through primary publication, synthesis, and instruction." Each one of these is an ideal and as such has been violated.

Young Michael Faraday was evidently impressed with the ideal ethic of science and resolved to enter the field. A bookbinder's apprentice, he attended the lectures of Humphry Davy at the Royal Institution and took copious notes. With a bound copy of his notes, he applied for a job with Davy. During Faraday's later years, a journalist asked him whether he had found happiness and a more decent atmosphere in science. Faraday replied that when he had quit business he thought he had been leaving the mean-spiritedness and pettiness of everyday life, but he found frail human nature the same everywhere.

More recently, a sociologist at Southern Illinois University studied more than 90 percent of Great Britain's two hundred and twenty nuclear physicists and found abundant distasteful situations. His report, published in a 1971 issue of *Minerva,* quoted one: "You must realize that the ethics usual in scholarly pursuits, literary and so on, do not apply to high energy physics. Outright dishonesty is prevalent, and there's not much stigma attached to being caught at it—I suppose the referee system is not too bad, but it also is definitely used immorally—to delay your competitor."

The failure to have the ideal ethic in practice is comparable to the relation between many scientific concepts and reality. Innumerable ideas in physical science are human inventions which the real world approaches but never reaches. In this class, among others, are perfect vacuum, ideal gas, center of mass, and uniformly accelerated motion in a straight line. Vacuums are common, but the one with absolutely nothing is a figment of our imagination; real gases subscribe only partially to the gas laws.

The failure to reach the ideal in ethics is parallel to the inability of scientific theories to encompass all facts within their purview. No matter what the theory, some anomaly exists to challenge the idea; when no more challenges arise, the theory soon becomes accepted as fact. The sphericity and rotation of the earth were long ago in the category of theory, but the presence of abundant positive evidence and absence of contrary evidence moved the concepts into the realm of undeniably true. The transition from theory to accepted fact is a surprise only because, as pointed out in Chapter 10, all parts of science are presented as having equal certainty when, in actuality, they do not.

The classification of theories into descriptive and explanatory categories may indicate that the descriptive encompass all the facts within its purview. Yet one of the best in the descriptives, the second law of thermodynamics, may be invalid for helium at exceedingly low temperatures. In such a circumstance, the immediate rationalization is that the law is not applicable or needs a small amendment.

The second law of thermodynamics is a favorite for analysis by philosophers. It is not as universal as a presupposition of science called "the principle of uniformity of nature," which is seldom discussed by scientists but often by philosophers. Despite the tremendous diversity, nature is considered to follow basic patterns, and this uniformity of nature is a basic faith of scientists—as much as many of them dislike the word "faith."

When judging the age of the earth through radioactivity, for example, the uniformity of nature must be believed in order to accept the result. Also taken as the truth are such doctrines as the rate of decay of the radioactivity is constant over time, the measuring instruments do not mislead, and no one has added or subtracted radioactive material from the sample. These factors are not usually considered actively although they may influence the result, and all of them depend upon the uniformity of nature for their acceptance as truth.

A fundamental assumption of science comparable to the principle of the uniformity of nature is the idea that man can solve problems. This modest claim is responsible for the continuation of the scientific enterprise, yet its misinterpretation has caused much anguish. Without the optimism that human beings are capable of knowing the architecture and processes of the external world, scientists would not even begin to work. When overextended, however, immediate answers are expected to every dilemma. It comes as a surprise—but should not—that science does not have answers to a great many questions.

SELECTED READINGS

Bronowski, J. *Science and Human Values* (New York: Messner, 1956).

Brumbaugh, Robert S. *Ancient Greek Gadgets and Machines* (New York: Crawell, 1966).

Glass, Bentley. *Science and Ethical Values* (Chapel Hill: University of North Carolina Press, 1965).

Goodfield, June. "The Tunnel of Eupalinus," *Scientific American,* **210** (June 1964), 104ff.

Hagstrom, W. O. *The Scientific Community* (New York: Basic Books, 1965).

Ravetz, Jerome R. *Scientific Knowledge and Its Social Problems* (New York: Oxford University Press, 1971).

Wade, Nicholas. "Special Virus Cancer Program: Travails of a Biological Moonshot," *Science,* **174** (December 24, 1971), 1306-1311.

PART FOUR

Misleadings

The public has come to accept the inevitability of misleading statements in newspapers and magazines as well as on radio and television. Some of the divergence from truth is excused as inherent in any communication; another rationale is that special pleading is a right of those who control the media. Books are also excused because the point of view is shown, directly or not.

A pervading myth is that science and the misleading are incompatible. Without formal public relations, science has developed a splendid image of omnipotent truth. Nonetheless, this section reveals some of the misleading material distributed by organizations of scientists as well as by individual scientists.

CHAPTER 13

Misleadings by
Organizations of Scientists

Some American advertisers have not hesitated to mislead the public by associating science with their product. Before the advent of the "ecology" upsurge in the early 1970's when some thinkers blamed science for pollution, mouthwash, cigarettes, deodorant, and feminine hygiene products were presented as being in league with white-smocked research-ers.

A different kind of exploitation in the name of science was done by the founder of the scheme called dianetics, L. Ron Hubbard. When that movement collapsed in the 1950's, he started scientology, a similar kind of cure-all for the personal problems of men, women, and children.

There is no science in scientology and none in Christian Science—the church did not intend to mislead people with its title. Reputable scientific organizations do not knowingly mislead, and intentional deceit is very seldom characteristic of scientific organizations. Physicists do not knowingly mislead by referring to the constant speed a falling body may attain as the *terminal velocity* as if it were the final speed. Astronomers call the brightness of a star at the standard distance of 32.6 light years its *absolute* or *intrinsic brightness,* although the impression of an absolute value is not intended. At the beginning of the twentieth century many American biologists did not knowingly mislead by being active in the eugenics movement. Their activity gave some scientific credence to the Immigration Restriction Act of 1924 as well as to the sterilization laws adopted by a majority of states.

The role of misleading comes to scientist groups through miscon-ceptions, momentum of events, or bureaucratic bungling. The role of women in science, for example, is usually not knowingly disparaged. Every western country, from Marie Curie in Poland and France to Caroline Herschel in England and Germany, has its feminine heroine of science. Yet the United States listing of working scientists has been called *American Men of Science.* This wrong will be righted. At its 1971 meeting, the Council of the American Association for the Advancement of Science adopted a motion calling attention to the lack of a central listing of women in science and to the fact that "the talents and contributions of women in science are not fully recognized." In 1972, only eleven women were among the more than nine hundred members of the National Academy of Science.

Depending on perspective, the prophets of doom or the optimists are misleading the public. Which group has the misconception? In March 1972, the Club of Rome, an informal international organization of seventy scientists, industrialists, and intellectuals, announced the results of an eighteen-month study, "The Limits of Growth," that indicated the collapse of civilization. Unlike similar dire predictions by economist Thomas R. Malthus in the nineteenth century, they pinpointed the middle of the next century as the time when pollution would be out of control and natural resources in short supply. On the other hand, those who encourage economic growth argue that the study is irresponsible nonsense with unwarranted assumptions and not enough data for extrapolations. The editor of the British magazine *The Economist* claimed new technology has always lifted man to new levels of achievement.

Varying conceptions, with one evidently misleading, are also involved in another vital issue. David D. Comey, director of environmental research for Businessmen for the Public Interest, charged early in 1972 that Atomic Energy Commission hearings on safety standards for nuclear power plants are "rigged." He said that all atomic generating stations should be shut down for modifications and new operating standards be set before a serious accident occurs.

Misconceptions about science have led scientists and nonscientists alike to self-deception and, consequently, to misleading others. Thus, the contention that scientific achievements are for the benefit of mankind is misleading; the error is the omission of the phrase "could be." Chlorine for water purification can also be a suffocating gas; dynamite for excavation can destroy buildings and kill innocent bystanders; bacteria are sometimes a friend of man. The unexamined assumption that scientific and technological achievements are always for the good is largely responsible for the environmental ravage of modern times. Almost any work in and out of science is capable of being used wisely or unwisely.

The first atomic bomb release over Hiroshima may be a classic case of man's inability to alter the buildup toward a predetermined goal. Leading scientists and some others in the atomic bomb project attempted to halt its use or at least have a demonstration for the Japanese and Germans on some uninhabited territory. Petitions to that effect were sent to Washington in 1945. However, plans for the bomb release over Japan went forward, and contrary opinions were largely ignored.

Bureaucratic bungling can also result in misleading the public as well as scientists. A recent illustrative case is about the efforts of the U.S. Atomic Energy Commission to find a place to bury radioactive debris. A salt mine underneath Lyons, Kansas, was tentatively chosen. During the period from 1955 to 1966, scientists from the Oak Ridge National

Laboratory studied the site. In March 1970, two Atomic Energy Commission representatives advised the governor of Kansas of the decision to use the abandoned mine and recommended he prepare the people for the event. In June, an AEC assistant general manager held a news conference in Topeka and announced the "tentative selection" of the Lyons site. A geology professor at the University of Kansas, William Hambleton, issued a challenging report in December 1971. He wrote: "As long as the waste containers maintain their integrity, only very small quantities of salt would be subject in high-energy, heavy-particle radiation. However, release might occur once or twice a year for about three years, and melting or explosion might cause containers to migrate to lower depths. . . . Accordingly, radioactive particles could migrate through the salt." Other features of the mine were found to be open to more questioning so that the AEC had to retreat from its position of tentative selection.

Organizations of scientists may mislead because they adopt the role of advocate rather than judge. They tenaciously cling to one idea, sorting evidence to support it and thwarting any challenges. Instead of an open mind for a multitude of hypotheses for phenomena, they display intolerant support for a favorite conception. Not only personal pride but also economics may motivate the unbending stance.

One of the sins of man, greed may also foment misleading. In 1967, Columbia University announced that it was undertaking the management of a new cigarette filter—the inventor had transferred most of his rights to the school, and the university made a great occasion of the event. Since the U.S. Surgeon General had reported a link between cancer and cigarettes a few years earlier, users of tobacco products were heartened by the announcement of a prestigious institution such as Columbia University. Smokers felt that the deleterious action of tar and other condensates would soon be a matter of history. However, the expectations were too high; the news conference was premature, and Columbia University soon voided its cigarette filter agreement; the public had been misled.

Columbia University was not the first nonprofit organization to mislead people about science and technology. More often the culprits have been scientific groups. One of the first groups in the United States was the American Society of Dental Surgeons. Their efforts at the beginning of the nineteenth century against the French Crawcours brothers who came to the United States to promote a new dental amalgam misled the public. The Crawcours were forced to leave the country, and the society passed a resolution declaring "the use of amalgam to be malpractice." When it tried to force members to pledge against using the amalgam under any circumstances, the society began to lose adherents and was disbanded.

On October 15, 1971, Senator Abraham Ribicoff of Connecticut announced serious charges against the Division of Biological Standards of the National Institutes of Health. He claimed the Division had permitted the potency of influenza vaccines to be labeled at greater than true values and had failed to insure the effectiveness of the vaccines. According to an eight-month study of the agency by the General Accounting Office (released March, 1972), more than half of the influenza vaccine lots certified over a three-year period were known to be less potent than the Division's own prescribed standard. The agency—now the Bureau of Biologics of the Food and Drug Administration—received another blow when a Philadelphia judge awarded a housewife more than two million dollars because the Sabin type III vaccine she had taken to prevent polio caused the disease in her. The judge commented that "the personnel of DBS charged with enforcing the regulation failed to read the regulation carefully or because they did not understand or failed to take seriously the strict duty imposed upon them by regulation. Either circumstance constitutes negligence."

High on the list of other offenders are the American Chemical Society and the American Medical Association.

When Rachel Carson's *Silent Spring* was published in 1962, the Manufacturing Chemists Association and its allies attacked the facts and contentions of the book with diatribe and innuendo. Jointly with the Nutrition Foundation, the Association issued a "Fact Kit" about *Silent Spring*. The kit contained a letter written by the president of the Nutrition Foundation, Dr. Charles Glen King, wherein the book was called unscientific and written by a "professional journalist—not a scientist in the field of her discussion." A chief ally was the American Chemical Society; the news columns and letters section of its *Chemical and Engineering News* contained much adverse criticism and relatively little support for Miss Carson's theme that manufactured chemicals were polluting the environment. Further, the magazine was not objective in reporting affairs in which the Pharmaceutical Manufacturers Association was involved. When the head of the journal's Washington News Bureau wrote an article in defense of the drug manufacturers and against federal regulation (December 19, 1966), many members of the American Chemical Society responded and the magazine published three protesting letters on January 23, 1967. One wrote: "The vicious attack . . . was an insult to the intelligence and integrity of the members of the ACS." Not the American Chemical Society but *Science News* published on April 22, 1967, the fact that the Department of Defense, a huge purchaser of drugs, turns down more than half of those offered to it. The manufacturers, selling to the American public, have either poor quality control or slovenly housekeeping.

The American Medical Association has probably the worst record for misleading its members and the public. Its intransigence against what was termed "socialized medicine" turned the association into a political pressure bloc, twisting facts and events to suit its end.

For most of its life from the founding in 1847 to World War I, the American Medical Association lived up to its constitutional aim "to promote the science and art of medicine and the betterment of public health." According to reporter Richard Harris, the AMA has since opposed "among other things, compulsory vaccination against smallpox and compulsory inoculation against diphtheria, the mandatory reporting of tuberculosis cases to public-health agencies, the establishment of public venereal-disease clinics and of Red Cross blood banks, federal grants for medical school construction and for scholarships for medical students, Blue Cross and other private health-insurance programs, and free centers for cancer diagnosis." Its fight against compulsory government health insurance lasted from 1920 through 1965, when Medicare was established. During the forty-five years, the AMA spent many millions of dollars and became a strong political force. In the process, false rumors were circulated, personalities suffered vituperation, courageous physicians were blackballed, doctor-patient relationships added a political dimension, and the AMA became immersed in advertising and propaganda.

In 1973, John Adriani, professor of surgery at Tulane University Medical School and a past chairman of the AMA Council on Drugs, testified before the Senate Monopoly subcommittee that the Council was abolished by the AMA to please the drug industry. He characterized the Council on Drugs as a group of independent thinkers whose compilation of drug evaluations could have brought the manufacturers to forgo advertising in AMA journals.

Misleading information and interpretations from the AMA was organized and directed, but occasionally zealous partisans issued their own material. The latter process is more frequent in other sciences. In at least three other cases, those of Velikovsky, the Loch Ness monster, and UFOs, the misleading sources were not organized but had the semblance of group action.

Velikovsky's book, *Worlds In Collision,* its publisher, and its supporters were harshly criticized by many scientists. There was no direct action by the American Geological Institute or the American Astronomical Association even though the leaders of the attack were specialists in the earth sciences. Perhaps the prestige of the leading opponents induced others to join in the condemnation of the library research; maybe the protesting scientists were appalled by theses at variance with what they had learned in schools.

The Loch Ness monster is no doubt promoted by the tourist bureau of Scotland as well as by the media. These two are responsible for highlighting misleading data although scientists have been involved.

The furor about UFOs (unidentified flying objects) has been stirred up by many unaffiliated would-be scientists, a very small number of established and reputable scientists, and the members of a small group claiming the objects are directed by extraterrestrial intelligence. Physicist Edward V. Condon was commissioned to investigate. His report, issued late in 1968, concluded that about 90% of all UFO reports are "quite plausibly related to ordinary objects." A "natural explanation" was proposed for only about half of fifty-nine cases but most of the remainder included inadequate or insufficient evidence or were impossible to evaluate for a variety of reasons. Nonetheless, others, including an astronomer or two, have cast doubt on the Condon report; the interested public must decide who, if anyone, is unintentionally misleading.

SELECTED READINGS

Dinsdale, Tim. *Monster Hunt* (Washington, D. C.: Acropolis Books, 1972).

Graham, Frank, Jr. *Since Silent Spring* (Boston: Houghton Mifflin, 1970).

Harris, Richard. "Annals of Legislation—Medicaire," *New Yorker* (July 2, 1966), 24-68; (July 9, 1966), 30-77; (July 16, 1966); 35-91; (July 23, 1966), 35-63.

_____.*A Sacred Trust* (New York: New American Library, 1966).

Juergens, Ralph E. "Minds in Chaos: A Recital of the Velikovsky Story," *American Behavioral Scientist* (September 1967) 4-17.

Lear, John. "Radioactive Ashes in the Kansas Salt Cellar," *Saturday Review* (February 19, 1972), 39-42.

Mackal, Roy P. "Sea Serpents and the Loch Ness Monster," *Oceanology International* (September/October 1967), 38ff.

Sagan, Carl, and Thornton Page (Eds.). *UFO's—A Scientific Debate* (Ithaca: Cornell University Press, 1973).

CHAPTER 14

Misleadings by Individual Scientists

At the end of the sixteenth century, John Dee, an English scholar, sincerely believed in astrology and alchemy. He associated with Edward Kelley who said he had obtained the secret of changing base metals into gold from a manuscript found in the ruins of an abbey. Both men demonstrated the conversion at the court of Rudolf II of Bohemia, who maintained a retinue of scholars as well as fakirs and magicians. Kelley hid his brother in the false bottom of a crucible, the brother substituted gold for the base metal in the kettle and the feat was accomplished.

Fraudulent claims comparable to the Kelley-Dee tableau have been made in more recent times. In 1812, one entrepreneur in the United States charged admission to observe an operating perpetual motion device. A commission of eminent engineers finally came to examine the complex machine. One of the commissioners brought his nine-year-old son, and the boy detected a flaw. He heard the noise of a turning crank although none was visible. Others tore away a wooden partition to expose an elderly man driving the apparatus. In 1898, another United States promoter was able to convince many people to invest in a perpetual motion device; this time, a clockwork mechanism was hidden in the base of the machine.

A more poignant drama unfolded around nineteenth-century French mathematiciam Michel Chasles. He had a penchant for collecting autographs and manuscripts, and an enterprising promoter sold him thousands of them. One purported to be correspondence between Isaac Newton, Blaise Pascal, and Robert Boyle wherein Pascal's priority in establishing the law of universal gravitation was established. Chasles presented the material before the French Academy of Sciences in 1867 and actively supported the proposition. Two years later, the perpetrator of the fraud was tried and convicted. Chasles testified that he had also purchased letters written by Galileo and Cleopatra—in French.

Twentieth-century scientists have been misled on numerous occasions. The culprits have been laymen or other scientists, and the subjects have been numerous, with psychology, geology and archaeology predominating. The explanations as far as can be ascertained have been the fun of playing practical jokes, dishonesty, economic gain, and unconscious deception. At the beginning of the century, the last was the reason why the horse Clever Hans was able to tap out arithmetical answers with a forefoot; almost any questioner knew the answers, wanted Hans to succeed, and unknowingly gave tiny cues such as very slight movements of the body.

One of the most famous cases of outright deception involved what was called Dawn Man, *Eoanthropus erectus*. The fossil bone found in England by Dawson was a genuine puzzler because the reconstructed specimen did not fit into the pattern of human development. The dilemma disappeared when it was shown that somebody had skillfully cemented together two different types of bone.

Similar affairs are common in paleontology, even with modern techniques of investigation. In 1911, a sea urchin fossil having teeth was found in a collection of Cretaceous fossils at the British Museum. No other toothed specimens were found among the many such fossils taken out of European chalk deposits. Only in 1969 was the problem solved when Porter M. Kier of the Smithsonian Institution—a man who had dissected many such fossils and found no teeth—looked closely at the one exception and saw what looked like dental cement. He used an abrasive air blast; the powdery material scattered; and the teeth, obviously implanted, came out easily. Those responsible for the sea urchin hoax cannot be identified because the British Museum has no record of when the sample was acquired and from what source.

In other cases, a misleader can be alleged. In biologist Paul Kammerer's laboratory in Vienna, for example, experiments purporting to show inheritance of acquired characteristics were performed. The finger of suspicion must move toward him and his associates. Likewise, the work done by Lysenko and his colleagues on the same topic, now discredited by some circles in the Soviet Union, made assignation of the error an easier task. Supporters of Kammerer and Lysenko categorize their work as genuine errors, not deception. Within the United States, a case came to light in 1973 when the Food and Drug Administration charged Dr. Wallace Rubin, an associate professor of medicine at Tulane University, with submitting false information to the government in 1967 and 1968. According to the agency, Rubin submitted "false reports" to Charles Pfizer and Company in connection with the drug benzquimamide and to duPont company in connection with metopimazine. The FDA said a similar incident with another researcher occurred in 1963.

When money is to be made in misleading scientists and laymen, the culprits are difficult to find. For example, generations ago, grave robbers broke into burial places and stole jewels and other valuables. Today, an illegal antiquities market is destroying and altering the physical remains of pre-Columbian civilization. One observer of the scene on the Yucatan Peninsula in Mexico wrote: "And the vandalism of pot hunters, who travel in large gangs and methodically destroy architecture in search of tombs and caches, is incredible. Hormiguero, which was until recently untouched, was exploited by such a gang in recent weeks—and now looks like a lunar landscape." As a step in the direction of stopping the pillage, in December 1971 Harvard University became one of the first to insist on proper title and pedigree for all acquisitions.

Those adept at writing proposals to acquire funds for research may be accused of misleading if their prospectus is not the truth. In 1972, Nobel prize winner Albert Szent-Györgi wrote that "all my fake projects were always accepted." Ridenour's story about his spectroscopist friend illustrates the same theme.

Partisans of philosophical views about science cannot be labeled as deceivers and misleaders; however, those terms are much used by their critics. Individuals must have the freedom and right to follow their intellectual conscience, else the house of science can fall asunder. Arguments between monists and dualists or mechanist and modern-day vitalists are as necessary for the well being and progress of science as are the subject-matter conflicts in Chapter 5.

The success of biologists, chemists, and physicists in unraveling the molecule of heredity has led to an increased voice for those viewing biological phenomena as eventually being described in terms of physics and chemistry. Such reductionisn has its opponents, but neither side can in any way be categorized as misleading. That indictment can be made only when any of the participants or bystanders write accounts for the public and generate the impression that one view is all wrong.

Many research scientists and science writers can be accused of misleading the nonscientist when presenting personal prejudice in philosophy, science, or public affairs as substantiated science. Accordingly, Jean Mayer, professor of nutrition at Harvard University, said at the American Association for the Advancement of Science meeting in December 1971: "While anybody has a right to capitalize on his fame as a scientist to discuss a matter about which he feels strongly, he should do so in, a way which doesn't confuse the public." The remark is simple to enunciate but its practical implementation is difficult. In the September 1, 1972, issue of *Science*, a letterwriter accused Mayer of doing exactly what he had decried. William M. Spicer of Atlanta, Georgia, wrote: "I disagree with a sentence in Jean Mayer's article 'Toward a National Nutrition Policy.' Mayer writes, 'as they became unneeded in agriculture, and at the same time, eligible to vote, an unspoken conspiracy of reactionary Southern officials created conditions such that the poor blacks would be driven to the North.'" Mayer was asked for evidence of the unspoken conspiracy.

When a scientist endorses a political candidate, condemns South Africa's racial policies, includes a chapter on God in his book dealing with popular astronomy or a chapter on cultural progress in his history of molecular genetics, the laymen can obviously discount what is outside the scientist's field of competence. It is easy to spot William Shockley, Nobel Laureate in physics, as a trespasser when he contends that Blacks have a lower intelligence because of heredity. However, ethnologists who give man a native factor of aggression or cultural analysts who wish to top the geological era with a new one indicating mental dominance and called "psychozoic" cannot be put off as easily.

Scientists may claim that the press and electronics media are guilty of the misleading, contending that they have been misquoted or their statements have been taken out of context. No doubt the accusations have substance, but the scientists are not always innocent. At a press conference on cancer research held at the National Academy of Sciences, October 1971, the scientists' statements had to be tempered by the journalists in order not to mislead the public. Again, after the 1973 American Cancer Society's seminar for science writers, one scientist said: "After listening to some of my 'colleagues' talk to some of the reporters, I'm never going to believe anyone who tells me he was misquoted if the press attributes some exaggerated claim to him. Of course, I'm not saying that this applies to everyone."

Reductionism and trespassing are involved in the efforts of some scientists to apply their conceptions of science and its procedures to social, political, and personal affairs. This brand of scientism misleads not only the young and innocent but also the professionals in the particular field, who may come to think that most scientists have the same views.

The identical culprits of reductionism and trespassing bring scientists into the quagmire of being seers for the culture of the future. On the basis of a promised technological feat, they pretend to foresee the automated kitchen, emotions completely controlled by mind, or travel through individual jet devices. The province of the technological future is not the sole property of the Sunday newspaper.

In 1973 at Birkbeck College, London, England, in his Bernal lecture, physicist Freeman J. Dyson of the Institute for Advanced Study in Princeton, New Jersey, one of the very few top-ranking physicists who never bothered to earn a PhD, made startling predictions in biology, not his field of competence. He foresaw the development of microorganisms to metabolize crude petroleum into desired products and to extract metals from the air and sea. Among his other visions were oysters extracting gold from seawater and organisms scavenging harmful radioactive by-products. When imagining closer to his own specialty, he mentioned self-producing machines colonizing parts of the solar system and the growth of trees on comets. At the symposium to celebrate the fiftieth anniversary of the Naval Research Laboratory, October 1973, he repeated some of his vision, emphasizing how the laws of economics could be outwitted by the construction of self-reproducing automatons.

Scientists, not publicists, have proclaimed the vision of having a completely understood universe, a contradition in terms with science as an endless quest. Physicist Albert Michelson said in 1892: "It seems probable that the grand underlying principles of physical science have been firmly established and that further advances are to be sought chiefly in the rigorous application of these principles to all phenomena . . . An eminent physicist has remarked that the future truths of physical science are to be looked for in the sixth place of decimals." Within a few years after this pronouncement, between 1895 and 1905, physics had a revolution with quantum and relativity theories, x-rays, electronics, and radioactivity. The scientist-predicter of a completely solved universe occurred again in 1970. In his retiring presidential address to the American Association for the Advancement of Science, biologist Bentley Glass said: "The great conceptions, the fundamental mechanisms, and the basic laws are now known. For all time to come these have been discovered here and now, in our own lifetime . . . we are like the explorers of a great continent who have penetrated to its margins in most points of the compass and have mapped the major mountain claims and rivers. There are still innumerable details to fill in, but the endless horizons no longer eixst."

Scientists who have authored biographies of other scientists have in essence misled readers with character descriptions that deify and glorify. All ugly facets of the individual are swept aside. Consequently, Newton is best understood through English literature specialist Louis T. More or historian Frank Manuel. The reluctance to present scientists as human beings with faults misleads the public.

Scientists are not angels, yet very few have been shown as less than perfect. Mathematician Jean Bernoulli, who lived at the end of the seventeenth and beginning of the eighteenth centuries, has been described as "violent, abusive, jealous, and when necessary, dishonest." Geologist John Woodward who lived at about the same time was "renowned for his eccentricities, pomposity, irritability and bad manners." Among others are Antonio Uliva, a member of the original society for experiment, arrested in Rome in 1667; Count Rumford, an eighteenth-century spy, sycophant, and acceptor of bribes; and Fredrick Accum, an early nineteenth-century German-born English chemist charged with stealing pages from and mutilating books at the Royal Institution Library. At the fourteenth annual meeting of the Society for the History of Technology in New York, 1971, a speaker cited American physicist Robert Millikan as having been involved in a dispute in 1918 wherein the inspector general of the Army recommended the surrender of Millikan's commission and his resignation from the National Research Council. In 1973, *Science* published an article accusing Newton of the fraud of fudging the data to fit his theory.

Galileo, Darwin, Einstein—indeed, and all the better-known scientists—need to be presented as human beings with failings. Their particular talents can be admired just as well in such a realistic context.

SELECTED READINGS

Coggins, Clemency. "Archeology and the Art Market," *Science,* 175 (January 21, 1972), 263-266.

Goran, Morris. *The Future of Science* (New York: Spartan Books, 1971), Chapter 13.

Koestler, Arthur. *The Case of the Midwife Toad* (New York: Random House, 1972).

Meyer, K. E. *The Plundered Past* (New York: Atheneum, 1973).

Williams, Stephen. "Ripping Off the Past," *Saturday Review,* 55 (September 30, 1972), 44-53.

INDEX

A

Abel, Sir Frederick 40
Academic Freedom Committee 39
Academic Marketplace, The 37
Académie des Sciences 3
Academy of Sciences, USSR 6
Accademia dei Lincei 46
Accum, Fredrick 115
Acheson, Edward 86
Adams, W.S. 91
Adelard of Bath 16
Adriani, John 109
Age of Reason 27, 32
Agricola 86
Akenside, Mark 51
Aldini, Giovanni 41
Alhazen 16
Allied Chemical Corporation 42
Allison, Samuel K. 9
Amateur Astronomers Association of New York 98
American Academy of Arts and Sciences 14
American Association for the Advancement of Science 19, 28, 93, 105, 113, 115
American Astronomical Association 109
American Chemical Society 20, 53, 56, 108
American Federation of Teachers 39
American Geological Institute 109
American Institute of Physics 7
American Medical Association 108, 109

American Medical Association Research Institute 25
American Physical Society 5, 43, 59, 93
American Physicists Association 12
American Physics Society 46
American Psychological Association 40, 80, 99
American Society of Dental Surgeons 107
American Society of Mechanical Engineers 12
Anaxagoras 31
Andromeda galaxy 79
Antiballistic missile (ABM) 37, 38
Antikythera machine 95
Anti-technology 26, 74
Applied Science and Technological Progress 70
Arago, Dominique 45
Arago, Francois 56
Archimedes 25, 63
Archytas of Tarentum 95
Aristarchus of Samos 97
Aristotle 15-17, 31, 32, 63, 68, 92
Armstrong, John 51
Arnold, Matthew 31
Ashby, Eric 74
Asimov, Isaac 52
Asterisks 98
Astin, Allen V. 4
Astrology 81
Astrophysical Journal 64
Atherosclerosis 41

Atomic Energy Commission 38,
71, 89, 106
Advisory Committee on
Reactor Safeguards 8
Atomic Industrial Forum 8
Avicenna 16

B

Bache, Alexander Dallas 56
Bacon, Francis 27, 31-34, 43, 62,
63
Bacon, Roger 16
Baltimore, David 9
Barnard, Edward 86
Bartlett, Dewey F. 56
Barzun, Jacques 59
Batavia, Illinois 23
Batelle Columbus Laboratories 70
Beekman, Isaac 97
Bell, Sir Charles 53
Bellman, Richard 55
Ben-David, Joseph 73
Bennett, Ivan 69
Bernoulli, Jean 115
Berthelot 92
Berthollet 56, 92
Berzelius, Jons. 86
Bigeleisen, Jacob 6
Biological Standards, Division of
108
Biological tinkering 29
Bioscience 52
Biot, Jean Baptiste 45
Birbeck College 114
Birth control 19
Bitter, Francis 60
Bloch, Herbert 12
Blumenbach, Johann 86
Boas, Franz 86
Bode, Henrik 69
Boerhave, Herman 18
Bohr, Niels 34, 55
Boltwood, Bertram B. 47
Boltzmann, Ludwig 47

Borek, Ernest 52
Born, Max 54, 62
Borst, Charles A. 40
Bosch, Carl 4
Boscovich 92
Boston Globe 38
Botanic Garden, The 51
Boyle, Robert 18, 87, 111
Brahe, Tycho 3, 13, 59, 72, 87
Breakthroughs in science 83
Bridgman, Percy W. 47, 89, 93
British Association for the Ad-
vancement of Science 45, 56
British Medical Journal 34
British Society for Social
Responsibility in Science 46
Bronk, Detlev W. 54
Bronowski, Jacob 86
Brooks, Harvey 72
Brown, Harrison 52
Brown, S. 52
Bryan, William Jennings 19
Buckley, William F. 26, 62
Bulletin of Atomic Scientists 37,
54, 67
Bunsen, Robert 89
Bureaucracy in science 96, 106,
107
Bureau of Biologics 108
Bush, Vannevar 54
Businessmen for the Public In-
terest 106

C

California State Board of
Education 19, 20
California Institute of Technology
25
Cambridge University 45
Campbell, Norman 72
Cancer 25, 33, 107
Special Virus Cancer
Program 96, 97
Cannizarro 56

Caplow, Theodore 37
Cardin, Jerome 86
Carlos I (Portugal) 10
Carnegie Commission on Higher
    Education 57
Carothers, Wallace 47
Carson, Rachel 108
Cassini 3
Cavendish, Henry 87
Cavendish Laboratory 45
Center for the Study of Science in
    the Public Interest 9, 62, 93
Chalidze, Valery 6
Chamberlain, Owen 39, 40
Changing ideas in science 61
Chasles, Michel 111
Chemical and Engineering News
    20, 26, 80, 81
Chemotherapy 29
Chlorophyll 78
Cholesterol 41
Christian Science 105
Christina of Sweden 13
Churchill, Winston 55
Citing prior accomplishment 98
Clairvoyance 33
Clausius, Rudolf 35
Clavius, Christopher 17
Cleopatra 111
Club of Rome 106
Coates, Robert 53
Cohen, Waldo 53
Coleridge, Samuel Taylor 51
Columbia University 47, 107
Columbus, Christopher 83
Comey, David D. 106
Committee for Environmental
    Information 8
Committee on the Life Sciences 37
Committee of Scientific Society
    Presidents 56
Committee on Social Policy 37
Commoner, Barry 37

Compton, Arthur 43
Compton effect 43
Comte de Buffon 17
Conant, James B. 46, 60
Condon, Edward V. 110
Confidence in human reason 100
Conklin, Edwin Grant 28
Cope, Edward Drinker 45
Copernicus, Nicolas 17, 43, 67, 86,
    87, 97
Corbino, Orso 64
Cornell University 42
Council of Europe 98
Council for a Livable World 93
Council for Science and Society 94
Coupling, J. J. 52
Crawcours brothers 107
Creationism 19
Creation Research Society 19
Curie, Marie 47, 87, 105
Curtis, Heber D. 41
Cuvier, Georges 18, 42, 64
Cyclamates 8
Cyclotron 98

D

Dalton, John 64
Daniels, George H. 98
Darrow, Clarence 7, 19
Darwin, Charles 1, 20, 34, 51, 63,
    85, 86, 116
Darwin, Erasmus 51
da Vinci, Leonardo 67, 92
Davy, Humphry 32, 45, 51, 64, 68,
    87, 88
Dawn Man 112
deBroglie, Louis Victor 86
Dee, John 111
DeGolyer, Everette Lee 47
Descartes, René 13, 31, 32, 87, 97
Desert Research Institute 57
Dewar, James 40
Dianetics 105

Dickens, Charles 24
Dictionary of Pure and Applied Chemistry 35
Diesel, Rudolf 55
Discovery by accident 89
Discovery of Neptune, The 96
Doctrine of Signatures 81
Dogmatism 33
Double Helix, The 57, 59
Drummond-Jackson vs British Medical Association 33
Duane, William 43
Dubos, René 52
Duerrenmatt, Friedrich 24
Duke University 98
Duplicating the tube 42
duPont Company 8, 112
Durrell, Lawrence 53
Dutch Reformed Church (South Africa) 20
duVigneaud, Vincent 52
Dyson, Freeman J. 114

E

Early technology 95
Eastman, George 86
Eban, Abba 13
Eccles, John C. 61
Economist, The 106
Edison, Thomas 49, 55, 86
Education Policies Commission 98
Education and the Spirit of Science 98
Ehrenfest, Paul 47
Ehrlich, Paul 37, 40
Einstein, Albert 43, 52, 59, 61, 63, 68, 92, 116
Eisenstein, Ferdinand 64
Electricity 41
Electromagnetic radiation 41
Emelyanov, A.S. 74
Energy conversion devices 46

Engelhardt, Vladimir 25
English science 34
Environmental Defense Fund 10
Environmental Protection Agency 8, 9, 90
Eoanthropus erectus 112
Eratosthenes 83
Ethyl Corporation 8
Eugenics movement 105
Evolution protest movement 19
Extrasensory perception (ESP) 33, 43
Eyepiece, The 98

F

Fairchild, Herman LeRoy 42
Faraday, Michael 34, 45, 47, 60, 64, 88, 90, 92
Federal Radiation Council 8
Federation of American Scientists 5, 11, 93
Ference, Michael 72
Fermi, Enrico 35, 46, 61, 64
Fields Medal 3
First International Congress of Chemists 13
First International Scientific Conference 13
Fischer, Emil 47
Fischer, Hans 47
Fleming, Sir John Ambrose 20
Flores, Nicolas Molina 5
Food and Drug Administration 9-11, 108, 112
Ford, Gerald R. 56
Ford Motor Company 72
Fourcroy, Antoine 63
Fox, William 85
Fracastoro, Girolamo 51
Frederick II of Hohenstaufen 11
Freedom of Information Act 9, 10
French, A.G. 60
French Academy of Sciences 42,

44, 111
Freud, Sigmund 86
Fritz London Memorial 98
Fuchs, Leonhard 86
Fuller, R. Buckminster 85
Funding science 3, 14, 15, 71, 75
G
Galen 83, 92
Galileo 17, 18, 25, 39, 49, 63, 64,
    85, 87, 92, 111, 116
Galileo Reappraised 67
Galvani, Luigi 40
Gay-Lussac, J.L. 44
General Electric Laboratories 85
Genetic engineering 20
Geocentric concept 79
Georgetown University 37
German science 4
Gesell, Arnold 86
Gesner, Konrad 86
Ghiselin, Michael T. 47
Gilbert, William 86
Gissing, George 23
Glacial erosion 42
Glass, Bentley 99, 115
Gofman, John W. 8
Goodyear, Charles 89
Gortori, Eli de 5
Graves, Robert 24
Greece 11
Green, David 43
Gregory, Richard 92
Grosser, Morton 96
Grosseteste, Robert 16
Gutbier, A. 47
H
Haber, Fritz 35, 55, 61, 68, 86
Haber, Norman 86
Haeckel, Ernst 53
Haensel, Vladimir 91
Hale, George Ellery 57
Halley, Edmond 34, 43
Hambleton, William 107

Hamilton, William Rowan 51
Hamilton College 40
Handler, Philip 6, 27, 57
Hannibal 25
Harvard College Observatory 41
Harvard University 6, 85, 93, 113
Haute-Provence Observatory 97
Health, Education and Welfare,
    Department of 10, 28, 42
Health Physics Society 42, 43
Heaviside, Oliver 86
Heisenberg's Principle 81
Heliocentric concept 17, 18, 20,
    43, 79, 97
Hench, Philip S. 57
Henry, Joseph 56
Heraclitus 31
Herbicides 8
Herschel, Caroline 105
Herschel, Sir John 34, 51
Hertz, Heinrich 41
Hertzsprung, Einar 86
Hevelius, Johannes 43, 92
Hexaemeron 15
Hines, Jerome 53
Hirohito 10
Hitler, Adolph 4, 24
Holloman, J. Herbert 70
Honigschmid, Otto 47
Hooke, Robert 34, 39, 43, 45, 87
Hosmer, Craig 7
Hoyle, Fred 46, 52, 74
Hubbard, L. Ron 105
Hubble, Edwin 86
Huggins, Charles 87
Hutton, James 41, 86
Huxley, Thomas Henry 20
Huyghens, Christian 13

I

Illinois Institute of Technology 70
Immigration Restriction Act of
    1924 105

121

Indiana University 98
Industrial Research 52
Institut Oswaldo Cruz 5
Institute of Electrical and Electronic Engineers 12
Institute for Enzyme Research 43
Institute of Society, Ethics, and the Life Sciences 37
Interdisciplinary Research Relevant to the Problems of Society 71
International Conference on Magnetic Resonance 96
International Conference on the Peaceful Uses of Atomic Energy 13
International Congress of Anthropological and Ethnological Sciences 57
International Congress of Mathematicians 3
International cooperation 12
International Geophysical Year 12
International Years of the Quiet Sun 12

J

Jena University 47
Jenner, Edward 92
Johns Hopkins University 69, 78
Johnson, Harry G. 69
Johnson, Lyndon 9, 71
Joshua 18
Joule, James 18
Journal of Experimental Psychology 40
Journal of Organic Chemistry 52

K

Kahn, Reuben L. 59
Kallman, Hartmut 4
Kamen, Martin 53

Kammerer, Paul 47, 112
Kant, Immanuel 32
Kapitsa, P.L. 64
Katchalsky, William 56
Keats, John 52
Keldysh, M.V. 6
Kelley, Edward 111
Kelvin 87
Kennedy, John F. 70
Kepler, Johannes 60, 72
Keynes, John Maynard 87
Kier, Porter M. 112
King, Charles Glen 108
Kistiakowsky, George B. 69
Klein, Felix 41
Krebiozen 33
Krutch, Joseph Wood 24, 25
Kyoto University 23

L

Laetrile 33
Lamarck, Jean 42, 64
Lavoisier, Antoine 6, 87, 92
Lawrence, Ernest Orlando 46, 97
Lawrence and Oppenheimer 46
Lawrence Radiation Laboratory 38, 39
Lawsuits in science 33, 39, 40
LeChatelier, Henri 61
Leclerc, Georges-Louis 17
Leibniz, Gottfried 68
Leith, Charles K. 55
Lennox, Edwin 86
Levich, Veniamin G. 5
Lick Observatory 41
Liebig, Justus 34, 45, 87
Life from lifeless 16
Life on other worlds 20
Life-span extension 33
Limits of Growth, The 106
Lindemann, Fredrick 46, 55
Little, Arthur D. 92
Loch Ness monster 109

122

Locke, John 32
Lodge, Sir Oliver 34
Lomonossov, Mikhail 51
London Times 33
"lone inventor" 85
Long, Franklin 11
LoSurdo, Antonino 46
Lunchtime politics 39
Luria, S.E. 47
Luther, Martin 18
Luyten, William J. 61
Lyége, Jean 51
Lysenko 5, 112

M

Mach, Ernst 41, 68
Magie, William Francis 59
Magnus, Albertus 16
Maimonides, Moses 16
Maizuru, Japan 23
Making of a Counter-Culture, The 25
Malinowski, Bronislaw 86
Malthus, Thomas R. 106
Mammon 24
Management and Control of Research and Development, The 73
Manchester University 45, 70
Mansfield amendment 71
Manuel, Frank 115
Manufacturing Chemists Association 108
Marcellus 25
March, Robert H. 93
Marcuse, Herbert 24
Maritain, Jacques 24
Marius, Simon of Anapach 39
Marker, Russell 88
Mars 78
Marsh, Othniel C. 45
Martin, James D. 56
Martinovics, Ignatius 6

Martyrs of science 92
Maslow, Abraham H. 78, 79
Massachusetts Institute of Technology 9, 38
Mather, Cotton 18, 95
Maupertius, Pierre de 45
Maury, Matthew F. 56
Maxwell, James Clerk 41, 47, 51, 52, 87
Mayer, Jean 113
Mayer, Robert 47
McCarthy, Joseph 38
McCormack, Mike 56
McGee, Reece 37
McGill University 25
McMillan, Edwin M. 38
Medawar, P.B. 80
Medicine as science 78, 109
Medvedev, Zhores A. 6
Mendel, Gregor 92
Mercury 60
Merton, Robert K. 64
Meyerhof, Otto 52
Michelson, Albert 115
Mill, John Stewart 32
Miller, Neal E. 80
Millikan, Robert 115
Minerva 38, 99
Mistakes in science 60, 61
Mohammed 16
Molecular genetics 29
Moon Illusion 98
Moon landings 7, 63
Moratorium for Science and Invention 25
Mordy, Wendell A. 57
More, Louis T. 115
Morgan, Lewis Henry 86
Morris, J. Anthony 46
Morse, Philip M. 38

N

Nader, Ralph 62

Nagata, Masaji 85
Napier, John 34
Napoleon 11
National Academy of Engineering 70, 71
National Academy of Sciences 6, 19, 27, 54, 57, 62, 69, 70, 71, 95, 114
National Advisory Commission on Health, Science and Society 37
National Bureau of Standards 4
National Cancer Institute 96
National Education Association 98
National Information System for Psychology 40
National Institutes of Health 7, 10, 108
National Radio Astronomy Observatory 73
National Research Council 8, 116
National Science Foundation 7, 11, 33, 70
Amendment Act of 1968 71
National Tuberculosis and Respiratory Disease Association 79
National Welfare Rights Organization 86
Nationalism in science 35
Natural selection, theory of 1
Nature 34, 51, 52, 57, 92
Naval Research Laboratory 114
Nebulium 90
Needham, Joseph T. 44, 86
Neptunists 41
Nernst, Walther 55
Newman Report 28
New Scientist 24, 57
Newton, Isaac 1, 11, 18, 27, 31, 34, 45, 47, 49, 51, 54, 63, 64, 68, 72, 85, 87, 92, 111, 115

New York State University 6
Nininger, Harvey H. 86
Nobel, Alfred 40
Nobel Prize 24, 40, 51, 64, 95, 98
Nordsieck, Arn 88
Northwestern University 11
Novum Organon 34
N rays 60
Nuclear power plants 8

O
Oak Ridge National Laboratory 106
Objectivity in science 25, 26, 62
Oceanographical Society of Japan 23
Ochoa, Severo 40
Office of Science and Technology 10, 69
Olbers, Heinrich 86
Oldenbourg, Henry 39
Operations Research 38
Operations Research Society of America 38
Oppenheimer, J. Robert 4, 46, 88
Oresme, Nicolas 16
Organic evolution 1, 6, 11, 20, 28, 78
Organization for Economic Cooperation and Development 73
Origin of Species, The 20
Origins of Theoretical Population Genetics, The 47
Ostwald, Wilhelm 41, 59
Ovishinsky, Stanford R. 46, 85

P
Parapsychology 33
Paré, Ambroise 51
Parliamentary Science Committee 92
Parmly, Eleazar 52
Pascal, Blaise 18, 68, 86, 87, 111

Pasteur, Louis 35, 44, 53, 89, 92
Perkin, William H. 89
Perpetual motion devices 42, 111
Perse, Saint-John 52
Pertet, Frank A. 86
Pesticides 8
Peter of Spain 16
Peters, Christian H.F. 40
Pharmaceutical Manufacturers Association 108
Pharmacology 81
Philip II (Spain) 34
Physics Today 52
Piccard, Jacques 26
Piccioni, Oreste 40
Pierce, John R. 52
Place, Marquis de la 11, 32
Planck, Max 4, 59, 61, 89
Plate tectonics 77, 91
Plato 15, 31, 32, 88
Plutonists 41
Poincare, Henri 28
Poisson, Simeon 87
Politics in science 56, 57
Pollard, E. 18
Pollution 37, 74, 105, 106
Pope John XXI 16
Pope Paul VI 19
Porta, John Baptista 51
Positivism 32
Precognition 33
Price, George R. 43
Priestley, Joseph 18, 87, 95
Primack, Joel 55
Princeton University 56
Prior training of scientists 86
Project Hindsight 70
Provine, William B. 47
Psychokinesis 33
Psychology of Science, The 78
Pugwash Conference on Science and World Affairs 14, 75

Pythagoras 31

R

Ramsay, Sir William 45, 46
Rankine, William 52
Reagan, Michael D. 73
Reber, Grote 64
Recognition of scientists 39
Redi, Francesco 44, 51
Reductionism 78, 113, 114
Reincarnation 33
Religious scientists 18
Restle, Frank 85, 98
Revolutions in science 91
Rhine, J.B. 43
Ribicoff, Abraham 108
Ridenour, Louis 67, 113
Right to acquire information 9
Rizzo, Patrick 98
Robbins, Frederick C. 57
Roberts, Arthur 52
Rochester University 6
Rockefeller University 80
Roszak, Theodore 25, 62
Rowland, Henry A. 69
Roy, William R. 7
Royal College of Surgeons 20
Royal Institution of England 68
Royal Institution Library 115
Royal Society 10, 23, 27, 32, 43, 45, 56
Royal Swedish Academy of Sciences 14
Rubin, Wallace 112
Ruckelshaus, William D. 9
Rudolf II (Bohemia) 3, 11, 13
Russell, Bertrand 26
Russell, H.N. 62
Russian science 6
Rutherford, Ernest 1, 25, 39, 40, 45, 59, 64, 83, 88, 89
Ryle, Sir Martin 46

S

Sagan, Carl 86
Saint Ambrose 15
Saint Augustine 15, 16
Saint Basil 15, 16
Saint Pierre, Bernarden de 23
Sakharov, Andrei 6
Salk Institute 86
Salviani 51
Saturday Review 25
Sautuola, Marquis de 91
Schaefer, Vincent J. 85
Schiaparelli 60
Schlesinger, James R. 71
Schoenheimer, Rudolf 47
Schrodinger, Erwin 62
Schwann, Theodor 44
Schwartz, Charles 39
Science
    bureaucracy in 96, 106, 107
    certainty in 100
    changing ideas in 61
    conceited 25
    defense of 28
    destructive 26
    development of 63
    entrepreneurs in 70
    ethics in 98, 99
    funding of 3, 14, 15, 71, 75
    ideals in 99
    immoral 24
    impractical 26
    lawsuits 33, 39, 40
    medicine 78, 109
    mistakes in 60, 61
    nationalism in 35
    objectivity 25, 26, 62
    politics in 56, 57
    revolutionary 25
    revolutions in 91
    sponsors of 64
    trespassing in 114
    unique personalities in 87, 88, 115
    view of the universe 26
    women in 105, 106
Science 20, 38, 42, 46, 55, 57, 61, 78, 96, 113
Science and Government 46
Science and religion forum 20
Science for the People 93
Science Journal 46
Science News 108
Science: The Glorious Entertainment 59
Scientific method 39
Scientific Research 43, 81
Scientism 54, 114
Scientists
    as advisors 55, 92
    as artists 51, 52, 54
    as brilliant 59
    as government administrators 10
    as lobbyists 12
    as politicians 56
    bachelors 87
    citing prior accomplishment 98
    cooperation 1, 12
    diligence 87
    martyrs 92
    prior training 86
    recognition 39
    religious 18
    self-educated 86
    socially responsible 8, 92-94
    suicides 47
    wealthy 54, 55
Scientists and Engineers for Social and Political Action 93
Scientists Institute for Public Information 93
Scientology 105

Scopes trial 7, 19
Segrè, Emilio 40, 61
Semmelweis, Ignatz 47
Senate Internal Security Sub-
    committee 6
Serber, Robert 88
Serendipity 89, 90
Shapiro, Howard 52
Shapley, Harlow 41, 88
Shockley, William 114
Silas Bent 23
Silent Spring 108
Sinsheiner, Robert 25
Sirius 91
Skinner, B. F. 52
Sloan-Kettering Cancer Center
    (NYC) 87
Slotin, Louis 92
Smale, Stephen 3
Smithsonian Institution 112
Snow, Sir Charles Percy 46, 52, 87
Society for Neuroscience 5
Society for Social Responsibility
    in Science 11, 93
Society of Surgeon Dentists 52
Solzhenitsyn, Alexander 24
Sommerfeld, Arnold 41
Sorbonne 35
Southern Illinois University 99
Spallanzani 44
Speer, Albert 24
Sperry, Elmer 86
Spicer, William M. 113
Spontaneous generation 44
Sprat, Thomas 32
Squaring of the circle 42
Stace, W. T. 26
Stahl, Georg 86
Stanford University 8-10, 55
Stellar parallax 43
Sternglass, Ernest J. 42
Struve, Otto 64
Subelectron 60

Sun-centered idea 18
Supersonic transport 8
Swiss National Science Foun-
    dation 12
Syphilis 40
Szent-Gyorgi, Albert 113
Szilard, Leo 93

T

Tait, Peter G. 35
Tamplin, Arthur R. 8
Tartaglia, Niccolo 34
Tax, Sol 57
Technology in Retrospect and
    Critical Events in Science 70
Telepathy 33
Texas Board of Education 19
Thales of Miletus 23, 67
Thenard, Louis 63
Thomas, Lewis 87
Thompson, Benjamin 57
Thompson, J. Eric 72
Thomson, J. J. 18, 55, 83
Thomson, William 51, 54
Tolstoy 26
Tower of the Winds 95
Toynbee, Arnold 75
Trisecting the angle 42
Tulane University 109, 112
Tunnel of Eupalinus 96

U

Ubaldo, Guido 63
Udall, Stewart 9
UFO's 33, 109, 110
Uliva, Antonio 115
Underground nuclear tests 8
Uniformity of nature 77, 78, 100
United Nations Conference on
    Food and Agriculture 13
United Nations Educational
    Scientific and Cultural
    Organization 14

University of Arizona 26
University of Birmingham 33
University of Bologna 41
University of Bonn 35
University of California 8, 38, 39, 47, 55, 56
University of Chicago 9, 38, 57, 69, 72
University of Houston 27
University of Illinois 52
University of Kansas 107
University of Maryland 53
University of Munich 47
University of Rochester 42
University of Virginia 73
University of Wisconsin 93
Urey, Harold 93
USSR Academy of Science 71

V

Vacuum tube 20
van't Hoff, Jacobus Henricus 51
Vauquelin, Louis 63
Velikovsky 109
Venus 60
Vesalius 83
Virchow, Rudolf 11, 18
Volta, Alessandra 40, 51, 87
Voltaire 18, 34, 45
von Behring, Emil 40
von Braun, Wernher 4
von Engeln, O.D. 42
von Hippel, Frank 55
von Humboldt, Friedrich Wilhelm Alexander 57
von Pettenkofer, Max 47
von Reichenbach, Karl 60

W

Waddington, C.H. 53

Wakeley, Sir Cecil 20
Waksman, Selman A. 40
Wallace, Alfred Russel 1
Walpole, Horace 34
Walter, W. Grey 52
Washington, George 49
Washington Post 32
Washington Star-News 19
Washington University 70
Watson, James D. 57, 59
Watson, James E. 46
Watt, James 70
Weaver, Warren 18, 53, 54
Weems, Parson 49
Weizmann, Chaim 56
Werner, Alfred 41
Whewell, William 32
White, Lynn Jr. 67
Whitehead, Alfred North 97
Whitehorn, John C. 78
Whitman, Walt 23
Whitney, Charles A. 88
Wideroe 97, 98
Wilberforce, Samuel 20
Wiley, George A. 86
Willstatter, Richard 64
Wilson, Woodrow 56
Winthrop, John 10, 95
Wohlstetter, Albert 38
Woodward, John 115
Wordsworth, William 51
Worlds in Collision 109
Wurtz, Adolphe 35

XYZ

Xerox Corporation 91
Yesenin-Volpin, Alexander 6
Zariski, Oscar 6
Zuckerman, Sir Solly 73